Under Southern Seas

The ecology of Australia's rocky reefs

edited by

NEIL ANDREW

Under Southern Seas

The ecology of Australia's rocky reefs

Krieger Publishing Company
Malabar, Florida

UNSW PRESS

Contents

for Grace

A UNSW Press book

Published by
University of New South Wales Press Ltd
and
Krieger Publishing Company
Exclusive distributor for: Americas (North, Central, & South), Caribbean, Europe, and Africa.
1725 Krieger Drive
Malabar, FL 32950-3323 USA
Tel: (407) 724-9542 Fax: (407) 951-3671
info@krieger-pub.com

First published 2000

National Library of Australia
Cataloguing-in-Publication entry:
Under Southern Seas: the ecology of Australia's rocky reefs.
Bibliography.
Includes index.
ISBN 0 86840 657 0. (UNSW Press)
1. Reefs—Australia.
2. Reef ecology—Australia.
3. Marine biology—Australia.
I. Andrew, Neil L. (Neil Leslie).

577.7890994

Library of Congress Cataloguing-In-Publication Data:
A catalog record for this book is available from the Library of Congress, Washington, DC.

ISBN 1-57524-141-2 (Krieger)

10 9 8 7 6 5 4 3 2

Designer Di Quick
Printer McPhersons, Melbourne

PHOTOGRAPHS
PAGE 1 Large Blacklip Abalone showing the clean shell margin that indicates recent growth, Waterfall Bay, Tasmania. *Simon Talbot*
TITLE PAGE Weedy Seadragons. Point Lonsdale, Victoria. *Mary Malloy*
CONTENTS PAGE In some parts of New South Wales, Cunjevoi can almost completely cover shallow reefs. Lennards Island, Eden. *Nokome Bentley*
ABOVE The Southern Keeled Octopus typically occurs on sand but forages over reefs at night. Rye, Victoria. *Mary Malloy*

Acknowledgments

The Fisheries Research and Development Corporation, the South Australian Research and Development Institute and New South Wales Fisheries were generous in their support of this book. From those organisations, Peter Dundas-Smith, Simon Prattley and Patrick Hone (FRDC), John Keesing (SARDI) and Bob Kearney, Rick Fletcher and John Glaister (NSW Fisheries) made things happen. I am grateful to the contributing authors for donating their time and expertise to this project so freely.

The professional photographers whose images bring this book to life were similarly generous in providing their photographs at heavily discounted rates. I am particularly grateful to Ken Hoppen and Rudie Kuiter for allowing me access to their extensive collections.

In addition to the contributing authors, comments by Neville Barrett, Dave Bellwood, Chris Battershill, John Booth, Paul Breen, Andy Davis, Doug Ferrell, Malcolm Francis, Harry Gorfine, Ken Graham, Geoff Liggins, Alistair MacDiarmid, Warwick Nash, Rick Officer, Larry Paul, Craig Sanderson, Andy Short, John Stevens, Fred Wells and Chris Woods improved the text. Thanks to Louise Egerton for her editorial expertise, and Di Quick, John Elliot and Nada Madjar from UNSW Press for making the book a reality.

Thanks to Sally McNeill for believing in the idea, seeing it through the difficult times and for her expertise in putting it all together.

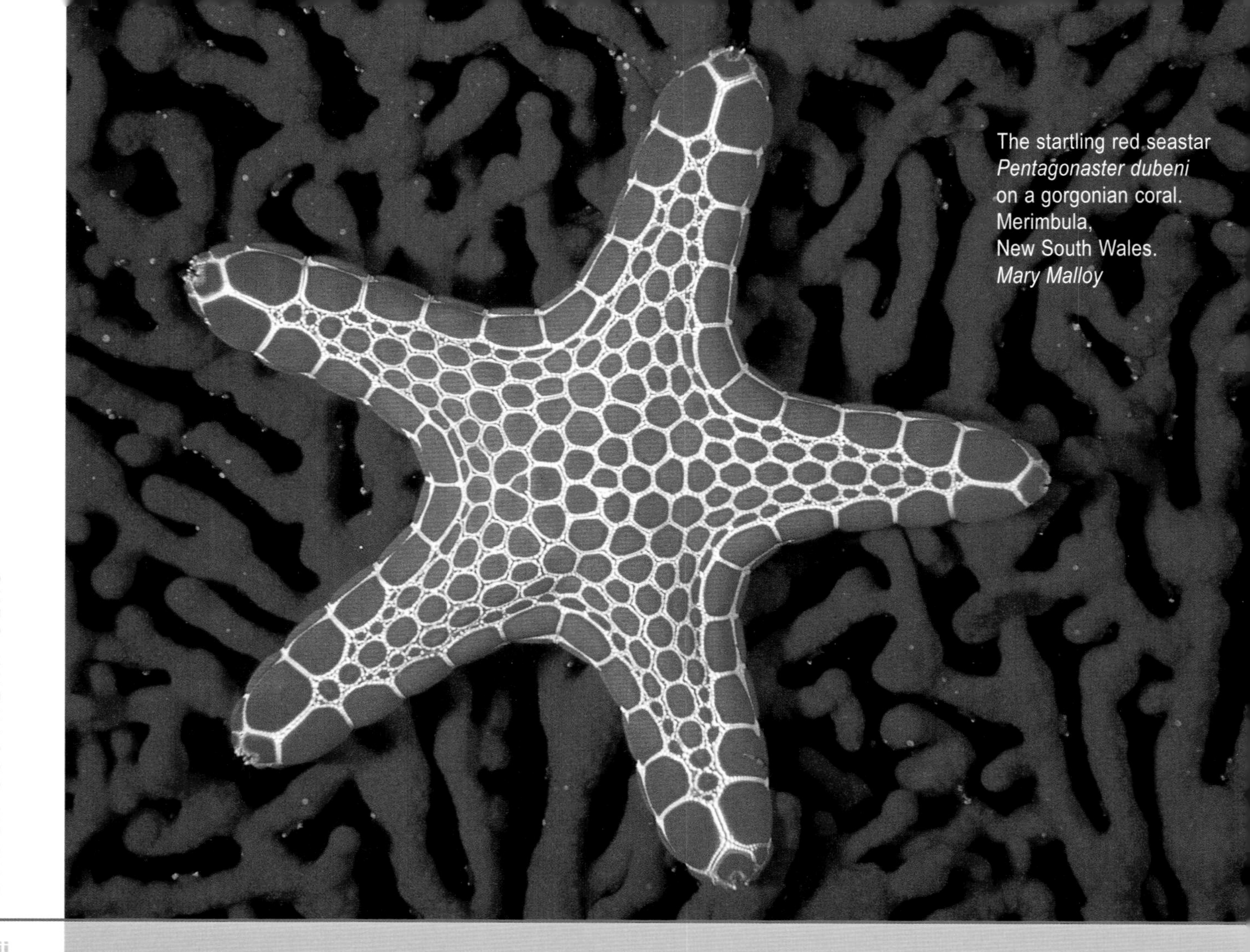

The startling red seastar *Pentagonaster dubeni* on a gorgonian coral. Merimbula, New South Wales. *Mary Malloy*

Contributing authors

NEIL ANDREW Principal Scientist, Shellfish, New South Wales Fisheries Research Institute, PO Box 21, Cronulla NSW 2230, Australia. *Present address* Project Director (Shellfish), National Institute of Water and Atmospheric Research, PO Box 14-901, Kilbirnie, Wellington, New Zealand

BARRY BRUCE Research Scientist, CSIRO Division of Marine Research, GPO Box 1538, Hobart, TAS 7001, Australia

MIKE CAPPO Experimental Scientist, Australian Institute of Marine Science, PMB 3, Townsville MC, QLD 4810, Australia

ANDREW CONSTABLE Senior Research Scientist, Australian Antarctic Division, Channel Highway, Kingston, TAS 7050, Australia

GRAHAM EDGAR Australian Research Fellow, Zoology Department, University of Tasmania, GPO Box 252-05, Hobart, TAS 7001, Australia

MATT EDMUNDS Senior Marine Ecologist, Consulting Environmental Engineers, PO Box 201, Richmond, VIC 3121, Australia

KAREN EDYVANE Scientist, South Australian Research and Development Institute, PO Box 120, Henley Beach, SA 5022, Australia

STEWART FRUSHER Senior Research Scientist, Tasmanian Aquaculture and Fisheries Institute, University of Tasmania, Nubeena Crescent, Taroona, TAS 7053, Australia

BRONWYN GILLANDERS ARC Postdoctoral Fellow, School of Biological Sciences, AO8, University of Sydney, NSW 2006, Australia

TIM GLASBY Postdoctoral Fellow, Centre for Research on Ecological Impacts of Coastal Cities, Marine Ecology Laboratories, A11, University of Sydney, NSW 2006, Australia

GARY HENRY Principal Scientist, Recreational Fisheries, New South Wales Fisheries Research Institute, PO Box 21, Cronulla, NSW 2230, Australia

BARRY HUTCHINS Curator of Fishes, Western Australian Museum, Francis Street, Perth, WA 6000, Australia

GEOFF JONES Reader, Department of Marine Biology, James Cook University, Townsville, QLD 4811, Australia

Gary Kendrick Lecturer, Botany Department, University of Western Australia, Nedlands, WA 6907, Australia

Mick Keough Reader, Department of Zoology, University of Melbourne, Parkville, VIC 3052, Australia

Michael Kingsford Senior Lecturer, School of Biological Sciences, AO8, University of Sydney, NSW 2006, Australia

Rudie Kuiter Honorary Research Scientist, Museum Victoria, 32 Mollison Street, Abbotsford, VIC 3067, Australia

Marcus Lincoln Smith Director, The Ecology Lab Pty Ltd, 25/28-34 Roseberry St, Balgowlah, NSW 2093, Australia

Michael Lowry Fisheries Manager, New South Wales Fisheries Research Institute, PO Box 21, Cronulla, NSW 2230, Australia

Paul McShane Director, Faculty of Fisheries and Marine Environment, Australian Maritime College, PO Box 986, Launceston, TAS 7250, Australia

Roy Melville-Smith Principal Research Scientist, Western Australian Marine Research Laboratories, PO Box 20, North Beach, WA 6020, Australia

Steven Montgomery Biologist, New South Wales Fisheries Research Institute, PO Box 21, Cronulla, NSW 2230, Australia

Mark Norman Research Fellow, Department of Marine Biology, James Cook University, Townsville, QLD 4811, Australia

Timothy O'Hara Research Scientist, Museum Victoria, 71 Victoria Crescent, Abbotsford, VIC 3067, Australia

Bruce Phillips Adjunct Professor, School of Environmental Biology, Curtin University of Technology, GPO Box U1987, Perth, WA 6845, Australia

Kylie Pitt PhD student, School of Biological Sciences, AO8, University of Sydney, NSW 2006, Australia

Jim Prescott Senior Scientist, South Australian Research and Development Institute, PO Box 120, Henley Beach, SA 5022, Australia

Danny Roberts Marine and Estuarine Ecologist, Wyong Shire Council, PO Box 20, Wyong, NSW 2259, Australia

Peter Shaughnessy Senior Principal Research Scientist, CSIRO Wildlife and Ecology, GPO Box 284, Canberra, ACT 2601, Australia

Scoresby Shepherd Senior Scientist, South Australian Research and Development Institute, PO Box 120, Henley Beach, SA 5022, Australia

Steve Smith Lecturer, School of Biological Sciences, University of New England, Armidale, NSW 2351, Australia

Peter Steinberg Senior Lecturer, School of Biological Sciences and Director, Centre for Biotechnology and Bio-innovation, University of New South Wales, Sydney, NSW 2052, Australia

Photographers

In addition to the authors, the following underwater photographers contributed slides:

Kelvin Aitken Kelvin Aitken Photography, 142 Keon Street, Thombury, VIC 3071, Australia

Neville Barrett Zoology Department, University of Tasmania, GPO Box 252-05, Hobart, TAS 7001, Australia

Paul Baumann Paul Baumann Photography, PO Box 182, Patterson Lakes, VIC 3197, Australia

Nokome Bentley Trophia Research and Consulting, PO Box 809, Claremont, WA 6010, Australia

David Bryant Sea Pics Photography, 35 Buldah Street, North Dandenong, VIC 3175, Australia

Bill Boyle 194 Station Street, Edithvale, VIC 3196, Australia

Peter Boyle 194 Station Street, Edithvale, VIC 3196, Australia

Neville Coleman Australasian Marine Photographic Index, PO Box 702, Springwood, QLD 4127, Australia

Rocky de Nys School of Biological Sciences and Centre for Biotechnology and Bio-innovation, University of New South Wales, Sydney, NSW 2052, Australia

Kevin Deacon Ocean Earth Images, 21A Pinduro Pl, Cromer Heights, NSW 2099, Australia

Ken Hoppen Ken Hoppen Photography, 20 Chelsea Rd, Chelsea, VIC 3196, Australia

Warren Jones Ocean Earth Images, 21A Pinduro Pl, Cromer Heights, NSW 2099, Australia

John Matthews New South Wales Fisheries Research Institute, PO Box 21, Cronulla, NSW 2230, Australia

Mary Molloy Nitrographics Underwater Photography, PO Box 1058, Altona Meadows, VIC 3028, Australia

Peter Morrison Sinclair Knight Merz Pty Ltd., PO Box H 615, Perth, WA 6001, Australia

Sue Morrison Western Australian Museum, Francis Street, Perth, WA 6000, Australia

Joachim Ngiam Atlantis Images, PO Box 1713, Hornsby Northgate, NSW 1635, Australia

Simon Talbot Zoology Department, University of Tasmania, GPO Box 252-05, Hobart, TAS 7001, Australia.

Pictilabrus viridus in kelp at Rottnest Island, Western Australia. *Barry Hutchins*

Foreword

Tim Winton

I grew up near the rocky reefs under southern seas. In fact, I suppose I could say that I grew up on those very reefs. I dived, surfed, fished and collected in the temperate waters of Western Australia and felt more at home when I was wet than when I was dry. The word reef usually puts people in mind of the coral reefs of the tropics — the vast profusion of colour, the immense diversity of species — and I suppose the tropics are still the glamour suburbs of the underwater world. But my neighbourhood, and the closest coastal marine environment to most Australians, is the cooler, more modest zone of the rocky reef.

Australians are lucky in that they are custodians of a coast of enormous length and variety. Our proximity to largely unspoiled seas and the pleasures they bring us is something we have taken for granted. The temperate coast has been our playground, our stress management course, our larder and means of turning a dollar for so long that it's only lately that any of us could even imagine the privilege slipping from our grasp. Until recently Australians probably didn't even appreciate their access to the living ocean as a privilege at all. It was simply a fact of life. It was certainly a central piece of mine.

As a boy I snorkelled the limestone reefs of the west coast to gather abalone, to spear octopus and blue groper. I hunted the West Australian Jewfish, the Baldchin Groper, the Harlequin, for food and for pleasure. Along the way I became acquainted with the huge tribe of wrasses. I learnt to appreciate sponges, soft corals, to identify rays and nudibranchs. I fished for snapper and squid, trapped crayfish and netted my share of prawns. Anything I learned about the sea was learned with an extractive tool in my hand — a spear, a net, a pot, a rod or gaff. Hunting does bring a kind of meditative focus to the mind, but it also funnels your attention in surprisingly narrow ways. It took me years to connect my hunting

△
Harlequin Fish are often seen resting on reefs in southwestern Australia.
Bald Island, Western Australia.
Barry Hutchins

behaviour with the effects it had on the rocky reef around me. The young, it seems, and the truly wilful, often suffer failures of imagination, and this was surely one. How did I let myself believe it would all go on forever? Well, we all thought it, I suppose. We thought in our traditionally small way, with our narrow focus and our firm belief that everything would always be all right.

Living in a whaling town for a time as a teenager shook me out of my comfort zone. Seeing sperm whales reduced to fertiliser shocked me. I got curious. Discovering how perilous the sperm whale's future was in the 1970s made me angry enough to see things differently. The sea I delighted in was not, after all infinite. It wasn't necessarily capable of absorbing our enthusiasm indefinitely. Australians are keen boaters and fierce anglers. We love seafood and work ourselves half to death to buy houses by the sea. All this activity takes a toll we're not in a position to assess with any real confidence. Despite our familiarity with the ocean our knowledge is very limited. But we do know that under-regulated fishing pressure causes the collapse of fisheries — we've seen it here and abroad. We know that land practices adversely affect our reefs and waterways. We see effluent and storm water outfalls changing water quality near our cities. More and more we are confronted by the fragility of what we have on our doorstep.

Little by little the sea-loving population grows more curious and slowly people are becoming more educated about what they take for granted. Grotesque overfishing and scandalously destructive coastal developments are not yet a thing of the past but they are at least hotly opposed and contested. Quotas and approvals are subject to sometimes bloody debate, surely a healthy sign of community interest. My fear is that until the level of education of ordinary boaters and divers and fishers gets up to speed, the power will naturally continue to reside with moneyed interests and the welfare of our reefs will be sacrificed. We live in an era of short-term pragmatism which doesn't make life easy for someone trying to preserve a habitat for its own sake. And it doesn't make it easy for a scientist studying the ecology of sea urchins. Unless, of course, the urchins can be exported. Neither does it make life simple for a commercial fisher tempted to press his quota. Or even his exporter who calls for a relaxing of limits even though he knows an increased catch is perilously unsustainable. Likewise it's tough on the recreational angler who must learn to accept smaller catches, and moreemission restrictions. But in time it will prove toughest on the children and grandchildren of all these people if we don't make hard decisions now and act with the conservatism that a precautionary approach demands. To the generations denied what we took for granted no special pleading will do. Their anger will be righteous enough. And the trail will likely lead back to many of us.

Renewed funds for research, informed pressure from ordinary people and the proliferation of marine protected areas as part of our overall management strategy are urgently required to safeguard the startling marine world we were born to. Works like *Under Southern Seas* can surely serve to convert our curiosity into knowledge. When I was a boy WJ Dakin's *Australian Seashores* was my constant handbook. Now at least we have something substantial to take with us beyond the surf line.

▽ The extraordinary Leafy Seadragon is endemic to southern Australia. *Mary Malloy*

Male Eastern Blue Groper
over reef, South West Rocks, New South Wales.
Joachim Ngiam

Introduction

Neil Andrew

Most Australians live south of latitude 30°S and within 100 km of the ocean. And almost all of us use the shallow rocky reefs that skirt this southern coast in one way or another — we swim, fish, go boating or use them indirectly by buying seafood harvested from them. Temperate subtidal rocky reefs support the most valuable commercial fisheries in the nation. Most important among these are the fisheries for abalone and rock lobsters. In the five southern states these fisheries were worth A$489 million in 1995-6 and accounted for 51 per cent of the total value of wild fisheries production in Australia (including the tropics). Considering each state separately, this proportion increases dramatically and was greatest in Tasmania and Victoria where abalone and rock lobsters provide 94 and 74 per cent of the total value of wild fisheries production respectively. Smaller commercial fisheries for reef fishes exist in all states. Other exploited taxa, such as sea urchins, octopuses, and even developing fisheries for jellyfish and kelp are partially or completely reliant on rocky reefs.

The magnitude of recreational fisheries on temperate rocky reefs are harder to quantify and statistics are largely unavailable. Available statistics suggest, however, that participation in these fisheries is high. For example, in 1984, 26 per cent of Australians went fishing at least once and 32 per cent of those fished from beaches or rocks. As another indication, 1.5 million residents in New South Wales fished at least once in 1995-96 and 82 per cent of this effort was in salt water.

Subtidal temperate reefs have been studied for many years, both to underpin management of fisheries and more generally to better understand the ecology of the reefs. These reefs are diverse, dynamic and capable of endless surprise. Often they are also cold, murky and difficult but their fascination remains and, for many people, they hold as much interest as the vaunted splendours of coral reefs. An important motivation in assembling this book has been to bring the hard-won knowledge gathered over more than 30 years of subtidal research on temperate rocky reefs to a wide audience. Since the pioneering days of the 1960s and early 1970s, there has been a enormous rise in subtidal research, both in the universities and government research agencies. Despite this effort, understanding of the ecology of subtidal reefs in southern Australia remains fragmentary. This book is therefore as much about what we do not know as it is about introducing the story so far. We have not attempted a compendium of everything that is known about shallow subtidal reefs, nor have we attempted to provide an identification guide to the thousands of species found. Instead, we have concentrated on broad patterns and the species and processes that are responsible for the way the reefs 'look' as you swim over them. Our synthesis is, of course, coloured by our own research interests and perspectives but we hope to convey a broad understanding of the ecology of rocky reefs in southern Australia. In doing so we hope that people will be motivated to fill the many gaps in understanding the ecology of subtidal rocky reefs.

NAMES AND CONCEPTS

We have used common names wherever possible but recognise that these may be vexatious given that they differ around the country. To avoid ambiguity, the scientific name is given the first time a common name is used in each chapter and thereafter if the possibility of confusion remains. Where the common name refers to one species only it is capitalised (eg, Snapper) and written in lower case where it refers to more than one species (eg, leatherjackets). The common name of several species of exploited fish differs from their established marketing name. Where this occurs, both are given.

The great majority of plants and animals in the sea, however, are not sufficiently familiar to be graced with a common name. These species are named in the text using their scientific names in the traditional binomial form of genus and species (eg, the large brown alga *Sargassum vestitum*). Where more than one species in the genus is mentioned this may be shortened (eg, *Sargassum vestitum* and *S. linearifolium*) and where more general statements are made, reference may be made only to the genus (eg, genus *Sargassum*) or a list of genera. In those cases where an organism has not been described or not identified to species level, the abbreviation 'sp.' is used (eg, *Caulospongia* sp.).

Some technical terms and concepts are unavoidable in a book such as this but we have endeavoured to keep them to a minimum and define them where they are first used. A recurrent concept in the book is that of a life cycle or a life history. This is the developmental journey an individual plant or animal takes through its life. Life in the sea has evolved an enormous diversity of life cycles, some of bewildering complexity that involve an individual passing through different forms and living in very different habitats. Two good examples are jellyfish and rock lobsters, both of which are described in later chapters. Often, there is a dispersive phase in the life cycle of marine plants and animals that begins as a fertilised egg drifting in the ocean. These eggs develop into forms called larvae if they are animals and propagules if they are algae. Both these terms are used extensively in the book. The sea provides a medium for transporting larvae and propagules long distances, making seemingly distant populations connected. This connectedness of populations and processes in the sea also means that impacts of man may be felt long distances from their source.

▷ Red Mullet are common in sheltered waters in southern Australia. Port Phillip Bay, Victoria.
Mary Malloy

ABOUT THIS BOOK

The first seven chapters of *Under Southern Seas* provide an introduction to large-scale patterns. These patterns are most obvious when the peculiarities and variability that characterise patterns on a smaller scale are averaged out. A description of the large ocean currents and processes that influence them is followed by overviews for each of the Australian states. At smaller scales, the 'noise' seen at geographic scales becomes all important and understanding the associations and relative abundances of species is crucial to an understanding of the ecology of individual reefs. A useful metaphor for these scales of observation are the differences in landscapes observable from satellite images or aerial photos and those from photographs taken on land.

The remainder of *Under Southern Seas* details present understanding of the main 'players' in these systems, such as large kelps, sea urchins and fishes. We also describe the ecology of abundant or high profile species, such as Grey Nurse Sharks, Snapper and Kingfish, lobsters and abalone.

MANAGEMENT OF TEMPERATE ROCKY REEFS

Although a discussion of the management of temperate rocky reefs is beyond this book, we hope the information contained will enhance the prospect of better management. Governments and their natural resource departments are increasingly singing the song of 'Ecologically Sustainable Development'. Although the phrase has eluded operational definition, the spirit is clear and appealing — we should use and conserve these resources to maintain their potential to renew themselves and to ensure that future generations have access to an undiminished resource. How this should be done is an ongoing and sometimes contentious issue. We hope this book contributes to the process of better managing these reefs by bringing an appreciation of the wonders and complexities of life on subtidal rocky reefs in southern Australia.

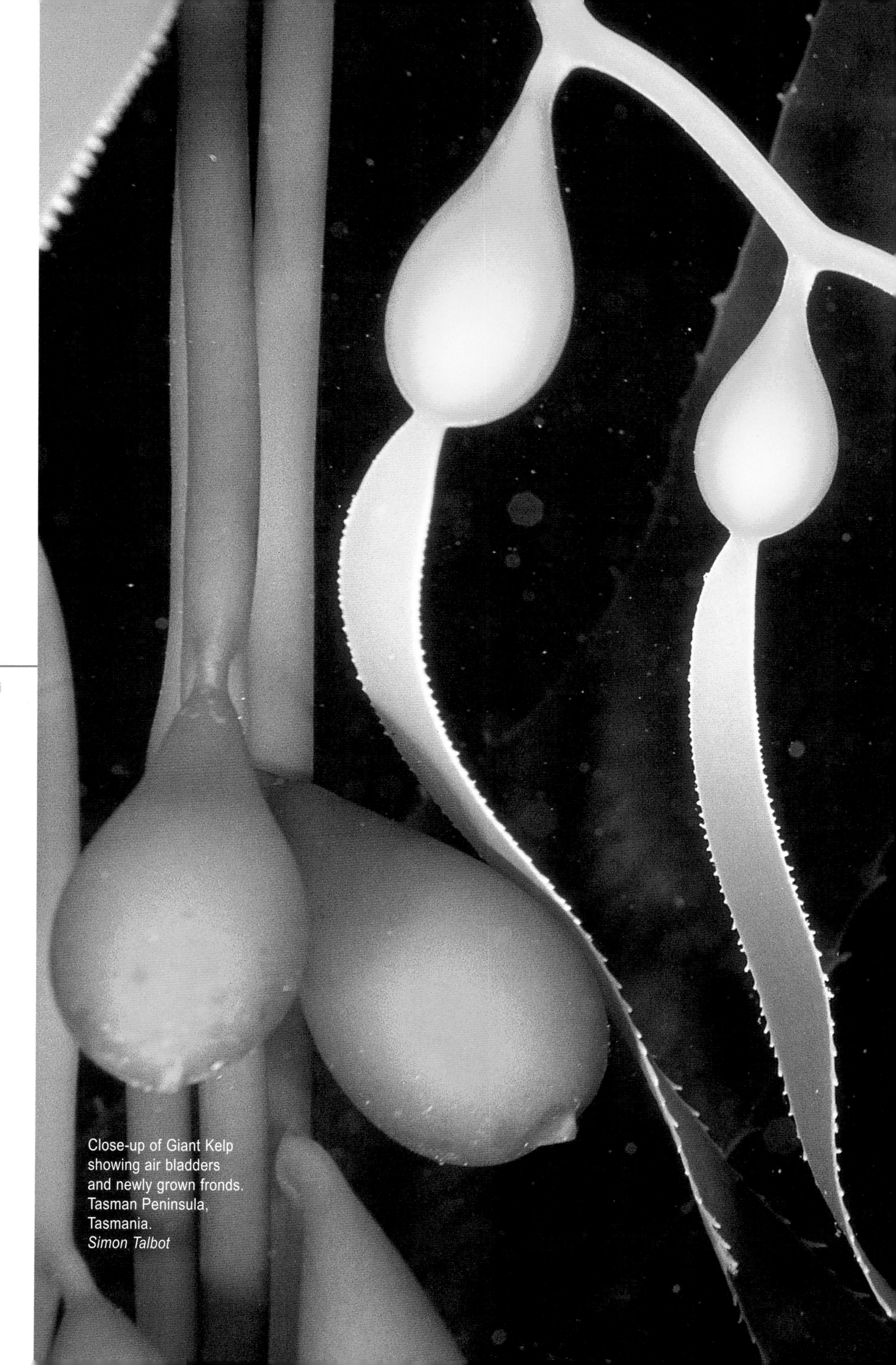

Close-up of Giant Kelp showing air bladders and newly grown fronds. Tasman Peninsula, Tasmania.
Simon Talbot

PART 1
Large-scale patterns

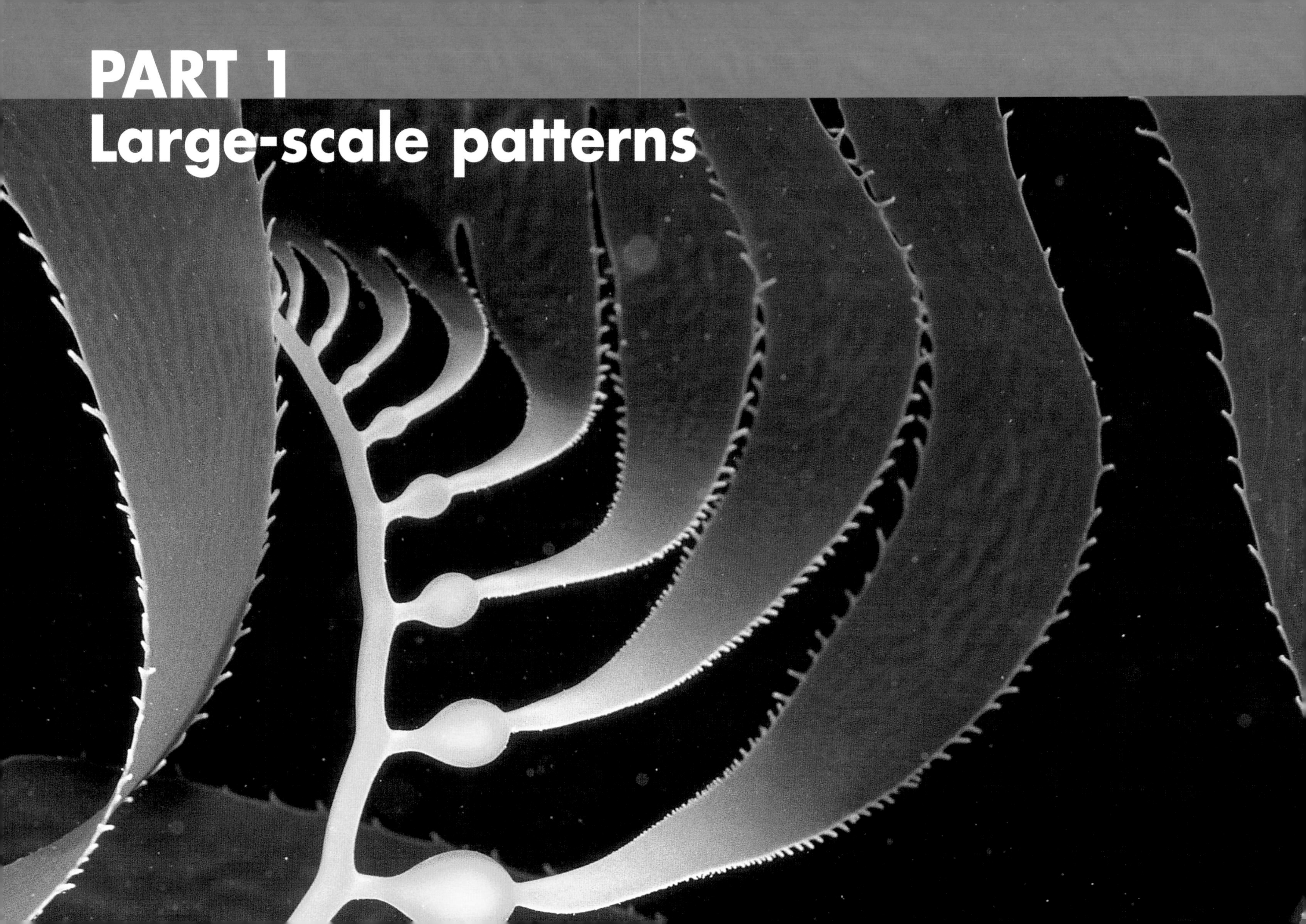

Plume of fresh water emerging from the Hawkesbury River, New South Wales. *Michael Kingsford.*

1 Oceanography and biogeography

Michael Kingsford

THE BIG PICTURE

Ocean currents are influenced by landmasses, global weather patterns, the rotation of the Earth and the gravitational influence of the moon. The most dominant flow in the oceans of the world is the west-wind drift, circling the globe south of latitude 40°S. Major weather systems generate huge swells in the Southern Ocean that travel around Antarctica undisturbed by landmasses. These cold waters support some of Australia's most productive fisheries in Tasmania and along the southern coast of the continent. The world's largest fisheries are generally found on continental shelves in less than 200 m of water and in areas where upwellings bring nutrient-rich water to the surface to nourish plankton and larger animals, such as fish and squid. The combination of these global processes with regional currents and the complex topography of the coast has fostered a rich flora and fauna on reefs in southern Australia. This diversity finds its greatest expression in the algae and sessile invertebrates, which are in many respects the most characteristic elements of the flora and fauna of southern Australia.

Many large-scale climatic and oceanographic processes have influenced the evolution of this flora and fauna, and still exert a large influence on the distribution of species. Most reef-associated algae and animals spend part of their lives in the plankton as larvae. Algal propagules and the larvae of fish and invertebrates spend a few minutes to many months in the plankton and most are weak swimmers for at least part of their early lives. Local currents may retain larvae near their natal reef, export them to distant but suitable habitat, or sweep them to certain death in the open ocean. Larvae are increasingly capable of swimming but may not attempt to control the direction and speed of movement until they detect a reef towards the end of the pelagic phase. Some fishes return to the location from which they were spawned, but others travel considerable distances in major currents. A dramatic example of this is the seasonal arrival of small tropical fishes to temperate latitudes on the east and west coast of Australia during summer and autumn. Except for hardy and well-defended individuals, such as lion fish and some wrasses, most die over winter, probably from stress due to cold and predation.

The species composition of plants and animals on a reef is the product of processes in the plankton, the nature of the reef and interactions with the organisms that live there. These influences are complex because many species, such as sea urchins or kelp, grow to become important in determining the biological structure of rocky reefs and in turn influence the types of animals and algae that settle and persist on the reefs.

On the east and west coasts of Australia, strong southerly flows bring warm water and larvae south to reefs that would otherwise be bathed in colder water In the west, the Leeuwin Current brings warm water to southern Western Australia and sometimes around to South Australia. In the east, the East Australian Current brings warm water from the Coral Sea down the New South Wales coast and sometimes as far south as the east coast of Tasmania. For eastern South Australia, Victoria and western Tasmania, however, these large currents have little impact. In these regions, storms, smaller tidal currents through Bass Strait, and localised upwellings, such as those near Robe in South Australia, have a greater influence.

MAJOR CURRENTS

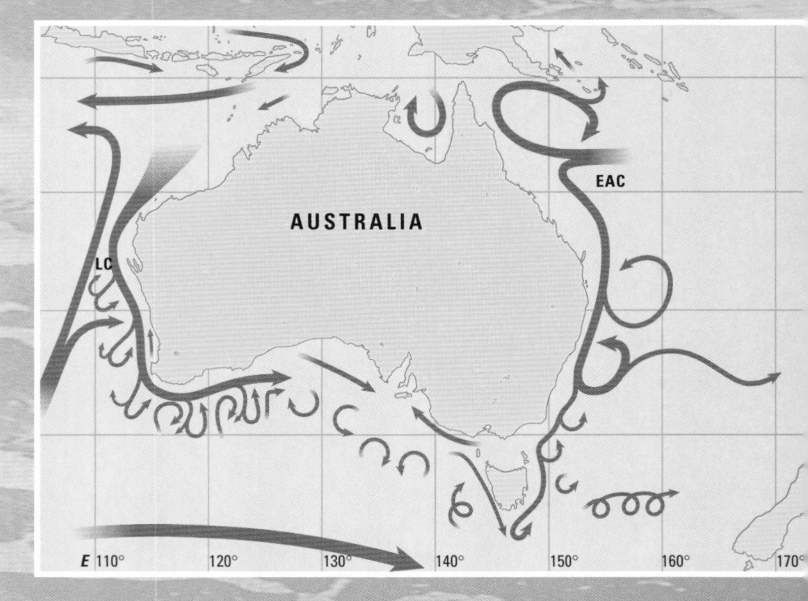

Map of Australia showing the major currents. Based on an original by George Creswell and redrawn with permission from Church and Craig,1998.
EAC — East Australian Current
LC— Leeuwin Current

THE EAST AUSTRALIAN CURRENT

When the East Equatorial Current in the Coral Sea meets the east coast of Australia around Lizard Island (latitude 15°S) it splits into a small northward current and a large southerly flow. The southerly portion moves along the length of the Great Barrier Reef, down the coast of Queensland and into the temperate latitudes of New South Wales as the East Australia Current (EAC). Along the New South Wales coast the EAC can reach speeds of 3 knots. The EAC is predictable as far south as Seal Rocks, after which its path becomes more erratic. As the current moves offshore it fractures into huge eddies of warm water up to several hundred kilometres across. These rotate in a counterclockwise direction as they spin off into the Tasman Sea or continue down the coast. In some cases they appear like doughnuts in satellite images because cool waters sometimes upwell in their centre. Warm eddies are often swept toward the shore for a few days before moving in a south-easterly direction. Swimmers in central New South Wales are familiar with the rapid changes in temperature during early summer as eddies from the EAC bring in warm water, followed by colder water several days later. Cool waters are upwelled near the coast in northerly winds.

The intensity and southern-most influence of the EAC changes with season. This warm current may reach Cape Howe or even Tasmania in autumn. The warm waters of the EAC are swept toward New Zealand. The transition between warm and cool waters in the Tasman Sea is known as the Tasman Front and is a conspicuous feature in temperature-sensitive satellite images. At some times of the year the EAC separates from the coast in the vicinity of Seal Rocks (latitude ~32°S). Sea conditions further south are influenced more by winds, swells generated by distant storms, tidal currents and the river outflows.

THE LEEUWIN CURRENT

Waters off the coast of Western Australia and to a lesser extent the Great Australian Bight are strongly influenced by the south-flowing,

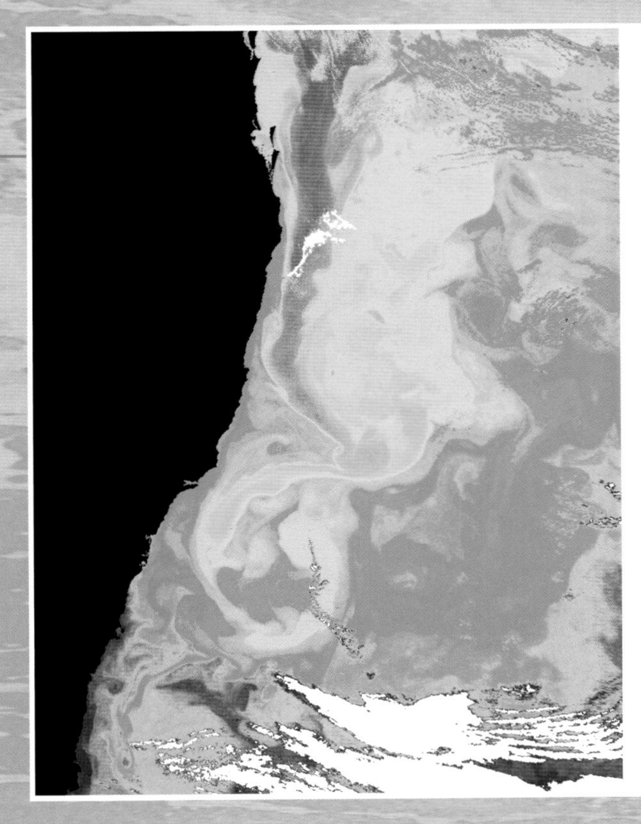

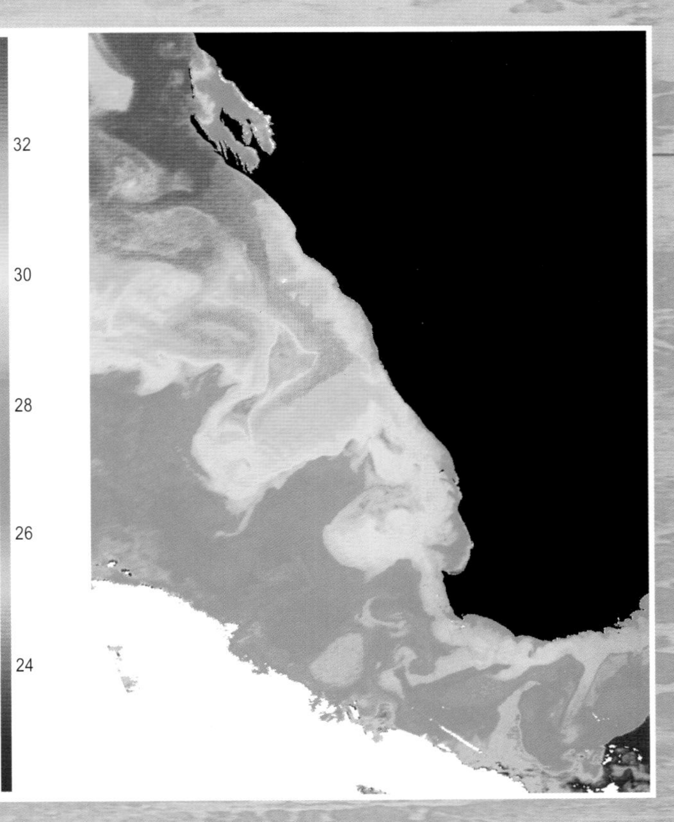

Satellite image taken on 29 September 1991 by AVHRR instruments on a NOAA weather satellite showing the EAC flowing down the east coast of Australia. Note the large eddies and changes in direction of the current. Data processed by the Remote Sensing Section of the CSIRO Division of Marine Research, Hobart.

Satellite image taken on 20 July 1997 by AVHRR instruments on a NOAA weather satellite showing the Leeuwin Current flowing down the Western Australian coast. Note the large eddies and changes in direction of the current. Data processed by the Remote Sensing Section of the CSIRO Division of Marine Research, Hobart.

warm Leeuwin Current. The Leeuwin Current streams over the continental shelf of Western Australia predominantly in autumn and winter. It travels 80–120 km/day across a narrow band of the continental shelf. This 'river in the sea' is 200 km wide in the north and narrows to 50–100 km in the south. Because it runs close to the coast, it forces colder northward currents offshore. No upwelling of nutrient-rich, colder waters occurs in Western Australia, so the warm inshore waters are relatively unproductive.

Although its influence is greatest on the west coast, the Leeuwin Current is also felt in the Great Australian Bight and can reach as far east as Port Lincoln and even Tasmania. The strength of the Leeuwin Current is influenced by patterns of water circulation across the Pacific and ENSO events (see page 6). When the ENSO cycle is not under strong El Niño conditions, the sea level is high in Indonesia, resulting in a strong flow of the Leeuwin Current. Under strong El Niño conditions, water transport across the Pacific from South America is minimal and the flow of the Leeuwin Current is relatively weak.

The Leeuwin Current is an exception in the southern hemisphere because it brings warm tropical water south along the western edge of a continent. In contrast, the major currents that influence the western shores of South America (Humboldt Current) and South Africa (Benguela Current) flow north, bringing nutrient-rich, cold water toward the equator. The Leeuwin also differs from other southward-flowing currents, which are common on eastern edges of continents in the southern hemisphere, because it flows most strongly during autumn and winter rather than summer.

The Western Rock Lobster (*Panulirus cygnus*) industry is worth over A$200 million/year. There is a strong relationship between the strength of the Leeuwin Current and successful recruitment of lobsters to shallow coastal environments. Lobster larvae can be transported up to 1500 km off the Western Australian coast before returning to settle on shallow limestone reefs. Meanders and jets of warm water from the Leeuwin Current move hundreds of kilometres into the Indian Ocean before returning to the continental shelf (see Chapter 14 for a more detailed description). Requirements of temperature, concentrations of food and transport in currents vary among species of larvae in Western Australia. In contrast to lobsters, scallops and pelagic fish such as Pilchards (*Sardinops neopilchardus*), show a negative relationship between number of larvae surviving and strength of the Leeuwin Current.

The Leeuwin Current disperses tropical fish, invertebrates and seaweeds southward into temperate waters during autumn and winter. It rounds Cape Leeuwin (latitude 34°S) and its eastward flow across the Great Australian Bight has some influence on the fauna and flora of that region. In Western Australia, mass spawning of corals coincides with the annual intensification of the Leeuwin Current, which transports larvae of many tropical species to Rottnest and Garden Islands, near Perth. A small persistent tropical component of the seaweed flora of Western Australia recruits into temperate waters from larvae carried south by the Leeuwin Current. Ten per cent (40 species) of algae found at Rottnest Island are tropical in origin.

ENSO AND GREENHOUSE EFFECT

The term 'El Niño' has become commonplace in discussions of the weather and its impacts on our lives. In Australia, El Niño weather conditions bring below-average rainfall, particularly in the latter half of the year. El Niño is part of a complex global pattern of climate that originates from interactions between the ocean and the atmosphere in the Pacific Ocean. El Niño conditions were initially thought to be anomalous, but are now understood to be one extreme in a large-scale phenomenon called ENSO (El Niño Southern Oscillation). At the other extreme, the conditions known as La Niña bring greater than average rainfall to Australia. These large-scale events have strong effects, so considerable research is directed at understanding ENSO, and in improving forecasting. A comprehensive outline of current understanding of ENSO is available in a book by Allan, Lindesay and Parker (*see* Further Reading, page 228). The authors have included a CD of satellite images with their book that dramatically illustrates ENSO phenomenon.

ENSO affects rocky reefs in several ways, especially through changes in the equatorial currents which influence the strength of the East Australian Current and the Leeuwin Current. These currents weaken during El Niño and strengthen during La Niña. Weather also influences marine ecology through storms and runoff of fresh water from cyclones. Large swells generated by cyclones sweep sediment from beaches and may bury adjacent reefs. Heavy swells may tear algae and sessile animals from reefs and deposit sediments. The above-average rainfall associated with La Niña years on the east coast of temperate Australia can dump heavily silt-laden fresh water onto coastal reefs, in turn affecting organisms on the reefs.

'Greenhouse Effect' refers to human-induced changes in the carbon dioxide content of the atmosphere and resulting warming of the planet. The effect of warming on coastal waters of Australia is potentially great. Warming may cause a rise in sea level through melting of the polar ice caps and thermal expansion (water becomes less dense as temperature increases). If sufficiently large, rises in sea level could change the nature of our coastline by drowning estuaries, lowlands and coastal communities. Currents around Australia could also change, for example areas of downwelling in the Great Australian Bight could become areas of upwelling — this could increase the productivity of the area and the value of fish stocks. Changes in currents could influence distribution patterns of marine organisms, including those associated with reefs. Species that rely on currents to return larvae to suitable habitat would be affected. Species with small distributional ranges could suffer irreversible declines as their environment changes. Currents, biogeography, commercial fisheries, aquaculture ventures and use of the land would change greatly in Australia as a result of significant global warming.

RIVERS AND PLUMES IN COASTAL WATERS

At a smaller spatial scale, river outflows have a major influence on the oceanography of coastal waters in many parts of Australia. Plumes of fresh water generated from these currents can extend kilometres from the mouth of rivers and estuaries, transporting sediment and nutrients from the land and the larvae of some species out to sea. In floods, the low salinity of surface waters kill algae and sessile and sedentary invertebrates (eg, ascidians and sea urchins) near the mouths of rivers. Some species of fish and invertebrates settle preferentially in estuaries — plumes may have an important role in providing a cue for larvae to find these habitats. Other fishes avoid areas of variable salinity and habitats with high sedimentary load. The impact of fresh water on nearby coastal reefs is a good example of the linkages between terrestrial ecosystems and the

△ Composite satellite image of Australia showing average winter temperatures. Data from the NASA Pathfinder data set processed by the Remote Sensing Section of the CSIRO Division of Marine Research, Hobart.

sea-changes in weather conditions on land can have a direct impact on ocean rocky reefs.

DIVERSITY AND ENDEMISM

While the marine flora and fauna of tropical Australia and the Indo-Pacific mixed some 20 million years ago (when the continental plates of Australia and Southeast Asia collided), the marine biota of southern temperate Australia has remained isolated for over 65 million years. Partially for this reason, southern Australian rocky reefs have a remarkably high degree of diversity and endemism. Other factors contributing to the uniqueness of this flora and fauna include the large continental landmass of Australia, particularly the extensive, though narrow continental shelf and the long east–west oriented, ice-free southern coastline (the longest stretch of south-facing coastline in the southern hemisphere).

The temperate waters from southwest Western Australia and along the southern coast of Australia to southern New South Wales and Tasmania are recognised as a major biogeographic region, the Flindersian Province. This province has some of the highest levels of species diversity and endemism in the world. This is true for the algae, bryozoans, ascidians and sponges, and mobile invertebrates, such as nudibranchs (sea slugs) and echinoderms (seastars, sea urchins and related taxa). In this region, approximately 600 species of fish, 110 species of echinoderms and 189 species of ascidians have been recorded. Of these, approximately 85 per cent of fish species, 95 per cent of molluscs and 90 per cent of echinoderms are endemic. In contrast, approximately 13, 10 and 13 per cent of fish, molluscs and echinoderms respectively, are endemic to the tropical regions of Australia.

The diversity and degree of endemism is possibly greatest in the algae. In the temperate regions of Australia there are an estimated 1155 species of algae, more than 70 per cent of which are unique to the region. These patterns are most extreme in the red algae; of the 658 genera (and 4000 species) which occur worldwide, approximately 43 per cent of the genera and 20 per cent of the species occur in southern Australia. Of these, roughly 800 species or 75 per cent are endemic. Approximately 57 per cent of the brown algae and 30 per cent of the greens are unique to the region.

UNUSUAL DISTRIBUTION PATTERNS

Currents may be responsible for some strange patterns of biogeography. Some fishes found in temperate and subtropical waters of Australia are also found around Japan, but not in tropical waters. For example, Golden-Spot Pigfish (*Bodianus perdito*), Comb Wrasse (*Coris picta*) and Stripeys (*Microcanthus strigatus*) are found both north and south of the tropics. These so-called 'antitropical' distributions have intrigued scientists for some time. Although movement of the Earth's plates might have caused these patterns, genetic evidence suggests that many taxa have not been separated for more than a million years, and most important geologic events relating to the movement of continents around Australia happened over 10 million years ago (eg, opening of the west-wind drift and the closing of the isthmus of Panama). Dispersal over the equator by cool trans-equatorial currents is the most popular explanation for antitropical distribution patterns. Waters of the tropical zone may have been transgressed in places by cool water during the ice ages of the last 1.5 million years or so. Temperatures would also have been cooler in deeper water in the tropics and this may have helped sharks and other taxa bridge the tropics. It is also possible that some groups could have dispersed between hemispheres via 'stepping-stones' of upwelling around northern Australia and Indonesia. Although there are many alternative hypotheses for these unusual patterns, the puzzle remains largely unsolved.

Aerial photograph of Green Cape, New South Wales, showing the predominance of Barrens Habitat (light-coloured reef outside the band of large brown algae extending from the intertidal zone). *BHP Land Technologies*

2 New South Wales

Neil Andrew

The underwater landscape of New South Wales' rocky reefs is as diverse and structured as that on land. The coastline spans nine degrees of latitude and a wide range of water temperatures and exposures to ocean swells from the Tasman Sea. In the north of the state, hard corals and kelp coexist at the Solitary Islands and on northern mainland reefs. On these same reefs the Crimson-banded Wrasse (*Notolabrus gymnogenus*) and the damselfishes *Parma unifasciata* and *P. polylepis* are common. In the far south, fishes such Blue-throated Wrasses (*Notolabrus tetricus*), Purple Wrasses (*N. fucicola*) and Banded Morwongs (*Cheilodactylus spectabilis*) are found with Bull Kelp (*Durvillaea potatorum*) and Crayweed (*Phyllospora comosa*) — species that are representative of cold-temperate regions. Between these extremes, many species characteristic of New South Wales are found along the length of its coast. Fishes such as the Eastern Blue Groper (*Achoerodus viridus*) and the Comb Wrasse (*Coris picta*) are abundant and well-known elements of the fauna. The distributions of many invertebrates, such as Eastern Rock Lobsters (*Jasus verreauxi*) and Black Sea Urchins (*Centrostephanus rodgersii*), are centred on New South Wales.

In this chapter I describe the major patterns in the abundance of algae and animals on rocky reefs in New South Wales. The focus of the chapter is well above the complexities of interactions among individual animals and algae and, except for some of the more important species, above the distributions of individual species. Present understanding of how individuals of species interact and why species occur where they do on reefs will be discussed in later chapters.

THE COAST

The New South Wales coast is typified by long beaches interspersed with estuaries and rocky headlands, particularly in the north of the

THE SOLITARY ISLANDS

BY STEVE SMITH

The coral *Acropora solitaryensis* was originally described from the Solitary Islands where it forms large tabular colonies on the sheltered sides of the islands. North West Solitary Island, New South Wales. *Steve Smith*

state. Rocky reefs account for only 30 per cent of the coastline but where found they differ enormously from the near vertical cliffs off Jervis Bay to the extended shallow-water reefs in southern New South Wales. North of Seal Rocks, subtidal rocky reefs are often small extensions of the intertidal zone that disappear into sand within 50 m. For much of the coast, however, subtidal rocky reefs extend more than 100 m from shore and cover many hectares. These reefs support a diverse mixture of algae, invertebrates and fish. South of Wonboyn, small nearshore reefs are again the predominant reef type, although extensive reefs exist offshore in deeper water near the Victorian border.

The ocean continuously shapes these reefs through the pounding of waves and the abrading influence of sand, pebbles and even grinding boulders. Large storm waves generate enormous hydraulic pressures which can tear fragments of reef away and move huge boulders, tens of metres in diameter, crushing all in their path. The way waves and suspended material shape rocky reefs varies among different types of rock. Depending on the softness of the rock and the force of the waves, rocky reefs can be weathered into smooth platforms with very little relief, or into complex structures such as bridges and caves. On deeper subtidal reefs, and those in sheltered waters, there may be a net deposition of sand and silt.

The geology of the central and southern coast can be divided into three distinct regions. North of Port Stephens, the rocks on the coast are sedimentary in origin. The east–west alignment of the geological faulting leads to the prominent headlands and long beaches characteristic of central New South Wales and predisposes the coast to a greater degree of sedimentation than further south. In the Sydney Basin, between Newcastle and Jervis Bay, the coast is mostly sandstone, which has weathered into

The Solitary Islands are a group of small, rugged islands off the north coast of New South Wales. There are five main islands in the group which are spread across the continental shelf between 2 and 10 km from the coast. Considerable attention has focused on these islands since the late 1960s when divers first provided descriptions of the coral-dominated communities and high diversity of species on the reefs surrounding the islands. Since 1991 the islands have been protected as part of the largest marine reserve in New South Wales. More recently, the 100 000 ha marine reserve was upgraded to the Solitary Islands Marine Park, the first marine park to be gazetted in New South Wales.

The Solitary Islands are a mix of the tropical flora and fauna of southern Queensland and the temperate species found in southern New South Wales. There are two currents that account for this mixing: the EAC described in the previous chapter, and a smaller inshore current that flows north. The influence of the EAC is most noticeable at the offshore islands, where corals can take up to 51 per cent of the available space on the reef. Corals are most abundant on the northern and north-western lees of the islands. More than 90 species of corals have been recorded from the region, most with tropical affinities and, for most of these, the Solitary Islands represents their southernmost record of distribution on the east coast of Australia. There are also a few corals which are endemic to the region (eg, *Acropora solitaryensis*), and several for which the Solitary Islands represents their northern limit of distribution (eg, *Coscinaraea mcneilli*).

Despite the high cover of coral, there is no evidence that the coral reefs are growing or accreting as classical coral reefs do. Rather, the corals grow as a veneer over the rocky substratum. A number of hypotheses have been tendered to explain this lack of accretion including: slow growth and low recruitment rates for corals; competition between corals and algae for available space; and the effects of periodic severe storms which remove corals from the substratum.

Within the Solitary Islands group, distance of the islands from the mainland has an important influence on the species found. Close to shore, rocky reefs are dominated by Common Kelp and sessile animals, particularly sponges and ascidians. Corals are scattered within the kelp forests. It is in these areas that interactions between corals and large brown algae are most obvious. Islands 2–4 km offshore support diverse coral communities and although other sessile animals and large brown algae are present, they are much less common than on the inshore islands. Coral diversity is at its highest at the two offshore islands (North and South Solitary Islands) where large brown algae are uncommon.

During the warmer summer months, and through into April, the islands support a diverse fish fauna which comprise permanent residents and also juvenile tropical species transported south by the EAC. Few of these recruits survive the cooler winter temperatures. An interesting feature of this summer influx of tropical species, is the difference among years in the mix of species found. This almost certainly reflects the type and number of tropical larvae which are present within the EAC at any one time, and also the duration of the period over which the islands are bathed in EAC waters.

sharp cliffs and flat intertidal platforms. Subtidally, these reefs typically have large undercut benches that are interspersed with large boulders and gutters. South of Jervis Bay, the coast is typically made of harder and older rocks, such as granite and gneiss. The main axis of orientation of the geology in this region is north–south, so longer stretches of rocky coastline occur without the interruption of beaches and estuaries, particularly in the far south of the state. Subtidal reefs in this region often slope sharply, with large vertical walls and boulders which give the reefs great structural complexity.

Over this template, the East Australian Current (EAC, see Chapter 1) exerts a pervasive influence on the distribution of many species. The EAC has a reasonably predictable but seasonally variable influence north of Seal Rocks, but breaks away from the coast into a series of eddies further south. Where the EAC leaves the coast and where the eddies touch on the coast further south are unpredictable. These eddies and associated counter currents have a major influence on water temperature and the transport of larvae along the coast. The EAC exaggerates differences in water temperature by bringing warm water further south than would otherwise be the case but relatively large differences in temperature remain. For example, in 1994–95 the sea-surface temperature at the Solitary Islands ranged between 19.4 and 27.3°C. In Eden in the far south, the average winter sea-surface temperature was 7°C lower than at the Solitary Islands and ranged between 12.3 and 23.7°C.

The distribution of many southern species ends in New South Wales. Locations of these limits vary but at several places the differences in the flora and fauna are more disjunct than others. Jervis Bay is an important point in the transition between warm temperate waters and the colder waters of the south. The largely oceanic waters of the bay support a high

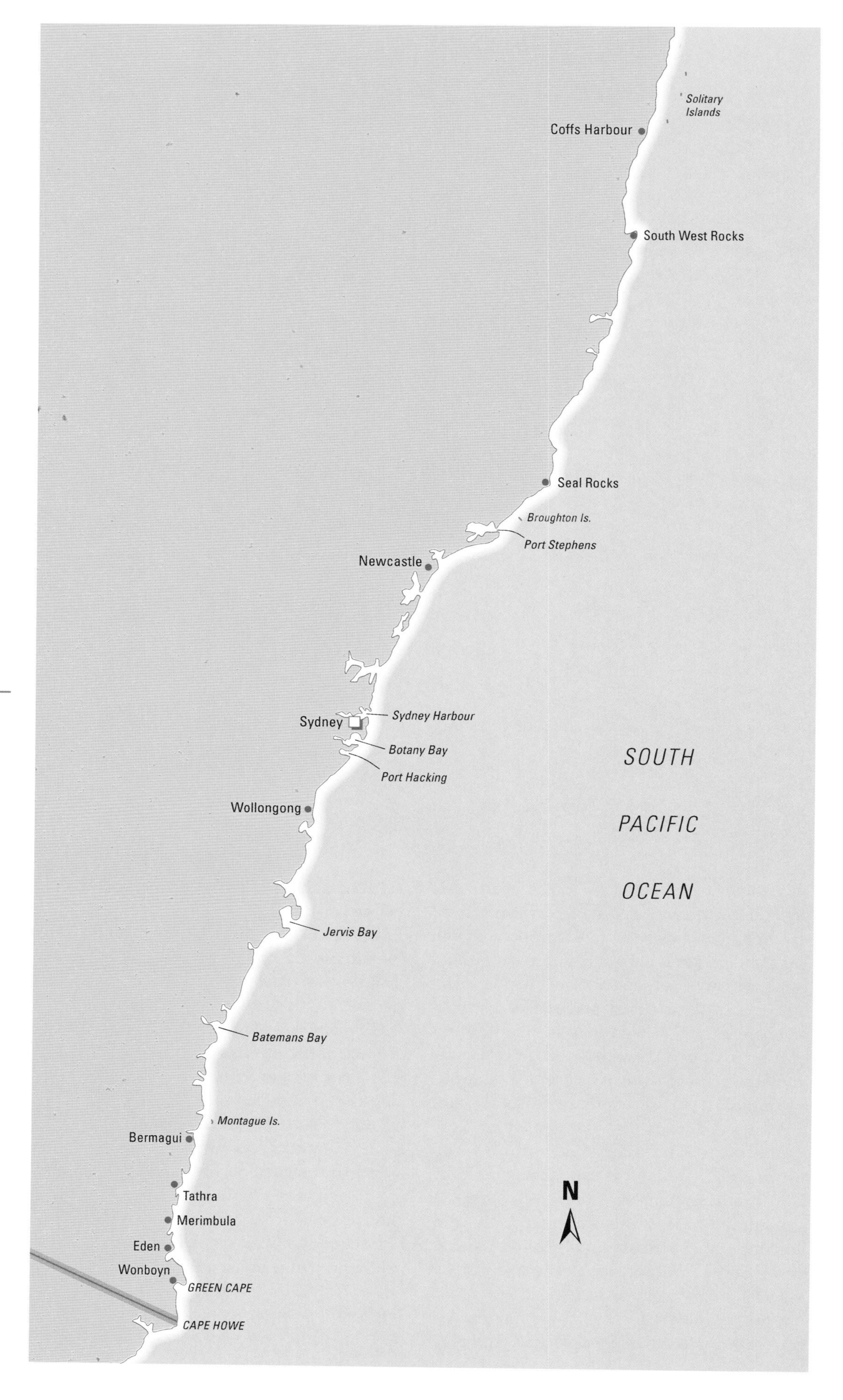

◁ Map of New South Wales showing place names used in the text.

△
Kelpfish are abundant on shallow reefs in New South Wales. Port Stephens. *Bill Boyle*

diversity of algae, reflecting this transition between the subtropics and the colder temperate waters. Of the nearly 500 species of algae recorded in New South Wales, nearly half (231) are found in Jervis Bay. Similarly, the reefs south of Green Cape, near the Victorian border are more typical of Victoria or Tasmania than the rest of New South Wales.

THE HABITATS

Within the broad limits introduced above there are consistent patterns in the relative abundance of invertebrates, fish and algae on subtidal reefs in New South Wales. These patterns are related to exposure to wave action, depth, and the effects of herbivores and the algae they eat. As on land, many species are commonly found together, often on relatively small scales. Much of the ecological research done on subtidal reefs in New South Wales has attempted to understand where and why these associations among species occur. A first step is to look for species groupings and to correlate them with features of the reef, such as depth and complexity. In New South Wales this exercise has revealed six such 'habitats'. Although they sometimes blend into each other and are difficult to define in other instances, this categorisation has proved to be a useful framework for studying the processes that make shallow subtidal reefs look the way they do.

In the first few metres of water on exposed and semi-exposed reefs in southern and central New South Wales, there exists a combination of species quite unlike that found in other parts of Australia. This habitat is known as the FRINGE HABITAT. Large brown algae such as the Common Kelp (*Ecklonia radiata*) and algae in the genera *Sargassum* and *Cystophora* are common. South of Sydney, Crayweed becomes increasingly abundant and south of Bermagui, Bull Kelp appears on the most exposed reefs. Although the mix of species in the Fringe Habitat may vary with latitude, a unifying feature of the habitat is that no single species of large brown algae dominates. These algae are interspersed with a large solitary ascidian, the Cunjevoi (*Pyura stolonifera*) and turfing red algae. Red and Purple Sea Urchins (*Heliocidaris tuberculata* and *H. erythrogramma*) often occur in shallow depressions carved in the rock and in deep cracks and crevices.

Single species from this mix of large brown algae and invertebrates can become dominant at some locations and form distinct habitats when they do. In the north of the state and at sites south of Merimbula and Wonboyn in the far south of New South Wales, Cunjevoi can occupy almost all available space and exclude most of the large brown algae. For unknown reasons this habitat, known as the PYURA HABITAT, is most common on small reefs. This habitat type, in which a solitary ascidian dominates subtidal reefs, is unusual and is only documented in New South Wales, central Chile and South Africa. In all instances it is formed by ascidians in the genus *Pyura* and probably the same species.

△
Dense patches of Bull Kelp are often found above *Phyllospora* Forests such as at Green Cape, New South Wales.
Nokome Bentley

◁ Fringe Habitat at Cape Banks, Sydney, New South Wales.
Neil Andrew

▷
Ecklonia Forests are found throughout New South Wales. Jervis Bay.
Kevin Deacon

South from Wollongong, Crayweed is increasingly the dominant large brown alga on exposed reefs. Dense forests of this species are recognised as a distinct habitat known as *PHYLLOSPORA* FORESTS. Relatively little is known about the ecology of this species, which is surprising given its prominence on reefs and its strong association with several important species, particularly Blacklip Abalone (*Haliotis rubra*) and Eastern Rock Lobsters (*Jasus verreauxi*) in New South Wales, Tasmania and Victoria. South of Wonboyn, the *Phyllospora* Forest habitat covers more than 50 per cent of nearshore reefs, but in New South Wales as a whole it covers less than 10 per cent of nearshore reefs.

South of Merimbula, Bull Kelp often forms a dense band on the shallowest margins of subtidal reefs (above patches of *Phyllospora* Forest). Although Bull Kelp is found as far north as Bermagui, it is less frequently found in dense bands. The continual abrasion of the reef by these plants allows little to survive under the canopy. Patches of Bull Kelp are known as the *DURVILLAEA* FOREST habitat. New South Wales is on the northern fringe of the distribution of this cold-water species and the plants are much smaller than their counterparts in Victoria and Tasmania, where plants can be very large and patches of Bull Kelp may cover large areas. In New South Wales, *Durvillaea* Forests cover less than 2 per cent of nearshore reefs and are patchy in their coverage.

Many species of fish are associated with these reefs in shallow water, the most abundant of which is the Rock Cale (*Crinodus lophodon*) which may be seen perched on the reef, often in groups. Little is known of the ecology of this ubiquitous species. In common with most locations in southern Australia, many wrasses inhabit rocky reefs in New South Wales, particularly in these brown algae habitats. Many of these wrasses have distributions centred in New South Wales, such as the Comb Wrasse, Crimson-banded Wrasse, and the Luculentus Wrasse (*Pseudolabrus luculentus*).

Anyone who has dived on rocky reefs in New South Wales will be familiar with the BARRENS HABITAT. These areas, which have no kelp or large algae, are found throughout the state, but particularly south of Newcastle. Between Sydney and Wonboyn, slightly more than 50 per cent of nearshore reef is Barrens Habitat and the percentage is higher (> 65 per cent) between Merimbula and Wonboyn. North of Sydney, the Barrens Habitat is less common and north of Coffs Harbour it is not the dominant habitat. South of Wonboyn, the nearshore reefs resemble those in Victoria — they are dominated by large brown algae, and Black Sea Urchins are relatively uncommon.

The most abundant algae in the Barrens Habitat are red algae known as crustose and turfing corallines. These algae grow as thin sheets or in short tufts and give the reef a characteristic pinkish hue. These algae are composed largely of calcium carbonate, which is resistant to grazing by sea urchins. On the surface of the coralline algae, many small limpets graze the films of diatoms and small ephemeral algae growing on the surface of the corallines. Limpets depend on grazing by the larger sea urchins to keep the substratum free of the larger turfing and foliose algae. When sea urchins are removed from areas of reef, turfing and large brown algae grow in profusion and limpets slowly disappear as they lose suitable surfaces to cling to. Sessile animals, such as sponges and ascidians, maintain space in patches of Barrens Habitat only where sea urchins do not graze heavily. Their cover rarely exceeds 15 per cent and they are mostly restricted to rock walls, often in association with the large barnacle *Austrobalanus imperator*.

There are generally fewer types of reef-associated fish in the Barrens Habitat compared to those with an abundance of large brown algae. White Ears (*Parma microlepis*, see Chapter 19) are the most abundant of the reef fishes in the Barrens Habitat. In areas with large boulders and many crevices, their territories are commonplace. Planktivorous fish may be abundant on the edges of reefs or in places where currents are strong.

Forests of Common Kelp cover the largest proportion of nearshore reefs aside from the Barrens Habitat. *ECKLONIA* FORESTS, as areas of this habitat are known, may cover many hectares and are most common within estuaries or sheltered bays, and in deeper water on exposed coasts. In sheltered waters, the stems or stipes of these plants are often long and woody and the canopy over the reef is dense. In depths less than 10 m on exposed reefs, individual plants are shorter and often battered by storms. *Ecklonia* Forests are broadly distributed in the state but are most common in central New South Wales. Around Port Stephens they cover approximately 35 per cent of nearshore reefs, but they decline in area on exposed reefs further south. South of Jervis Bay, *Ecklonia* Forests account for less than 10 per cent of the area of nearshore reefs and their relative abundance in sheltered bays and in

▷ The Barrens Habitat covers extensive areas of reef in New South Wales. Eden.
Ken Hoppen

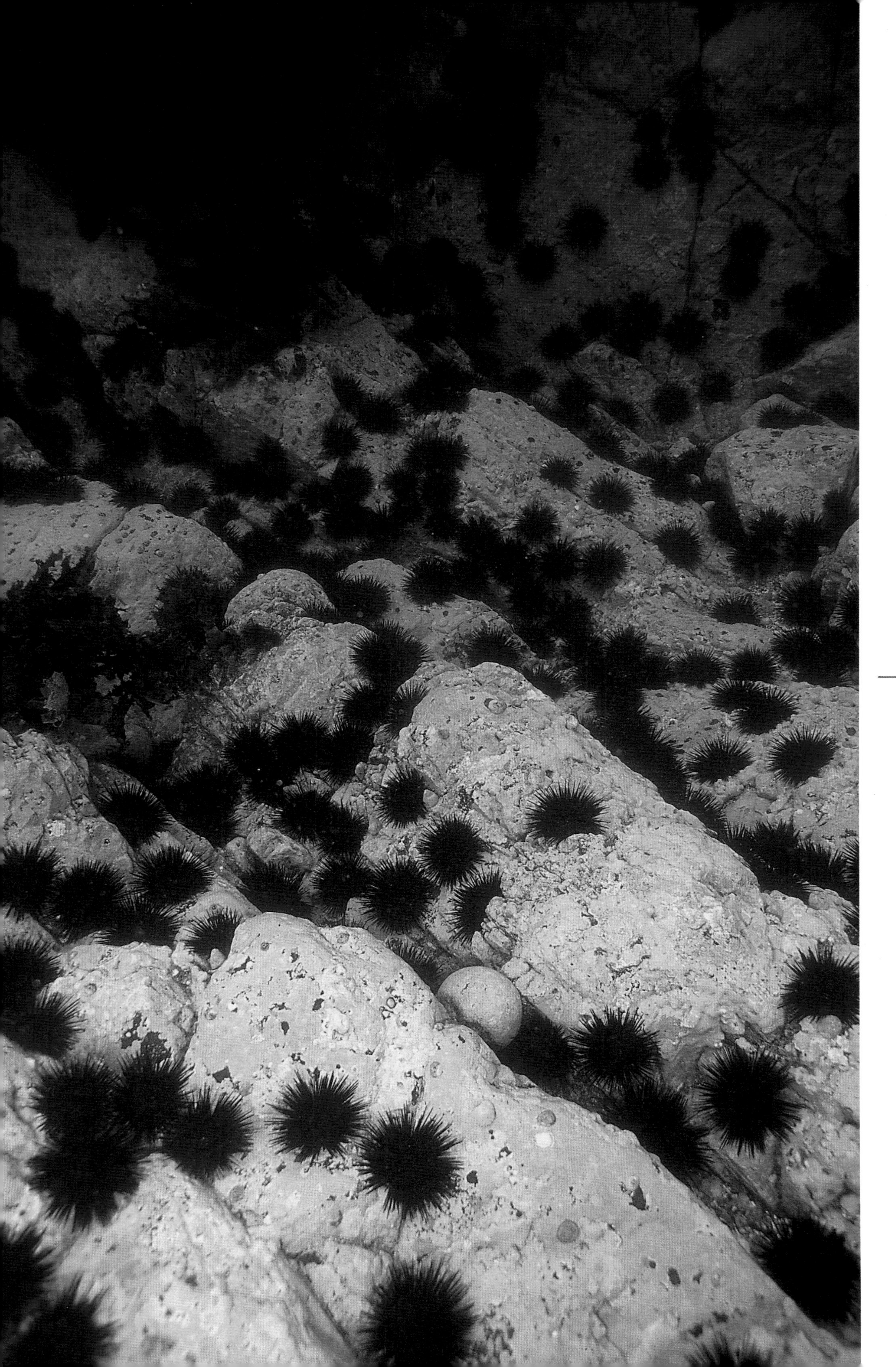

ROCKY REEFS BEYOND MOST DIVERS

BY DANNY ROBERTS

Reefs below the depth limits of most divers are the domain of sessile animals and encrusting algae. The sessile organisms found on reefs in water deeper than 30 m are mostly sponges, ascidians, cnidarians, bryozoans (see Chapter 16 for a detailed description of these types of animal), and encrusting algae. A first step in understanding these communities has been to describe patterns of diversity and abundance using remote controlled cameras and video equipment. Such studies completed off Sydney have revealed diverse but undescribed flora and fauna. About 65 per cent of the species of sponge collected from the deep reefs off Sydney have been new to science.

Many deep reefs are elevated above the surrounding sediment and act like islands. These reefs range in size from small outcrops and boulders to large flat shelves of many hectares. Some of the smaller reefs can be completely covered by sand and silt during storms. This silt matrix is a dynamic component of the deeper reefs and in some places accounts for a major proportion of primary cover. The areas of reef covered by silt generally increase with depth and changes are generally correlated with disturbance by large storms.

◁
Sponges (mostly genus *Mycale*) at 50 m.
Sydney, New South Wales.
Danny Roberts

The sessile animals found on these reefs have both erect and encrusting forms, and may cover up to 10 per cent of the reef. Encrusting red algae are generally the greatest occupiers of primary space, and only those species that can tolerate the low-light conditions found at 40 m depth can survive. The light at these depths is less than 1 per cent of that at the surface and these algae contain pigments that allow them to remain photosynthetically active. In some places these algae cover up to 90 per cent of the reef. The coverage of encrusting red algae declines with depth, probably because of increased siltation and the limited light.

Understanding these communities has been hampered by the fact that many species cannot be identified easily. The colour and growth form of sponges are affected by physical conditions, so these characteristics are unreliable for field identification. For example, *Spirastrella* sp. can grow as sheets on the reef in shallow water but has a massive structure in deep water. This difference may be related to differences in the physical environment such as currents, wave action and siltation. The colour of some sponges can also change with depth. As do many sessile animals from tropical reefs, a number of temperate reef sponges have symbiotic relationships with algae. These algae produce pigments which alter the colour of the external sponge tissue. With variable light, the types of symbiotic algae present can also vary. *Cymbastela concentrica*, a common cup-shaped sponge species, is found along the coast on shallow and deep reefs and on reefs in the mouths of Botany Bay and Broken Bay. The colour of individuals of this species reflects the amount of algae living within its surface tissue. For some temperate reef sponges, these algae contribute significantly to the overall nutrition of the sponge.

Algae and animals that live on shallow subtidal reefs close to large human populations are much more vulnerable to damage than the assemblages that live in deeper water. The construction and long-term operation of deep-water ocean sewer outfalls, however, represents a significant source of potential disturbance to these habitats. The response of sessile animals living on these reefs depends on the quality and quantity of the discharged sewage. Responses range from minor to large, rapid changes in assemblages. By their nature, these communities are difficult to study and the long-term responses of these communities to disturbance are largely undescribed.

water deeper than 10 m is more pronounced.

Ecklonia Forests generally are stable within their boundaries but undergo large fluctuations in the density of plants within them. These dynamics are discussed in detail in Chapter 7. Many fishes are associated with *Ecklonia* Forests, particularly the herbivorous Herring Cale (*Odax cyanomelas*) and the related Rainbow Fish (*Odax acroptilus*) which feed on small animals on the fronds of the kelp. Black Sea Urchins are rare in *Ecklonia* Forests but other smaller sea urchins may be locally abundant. For example, the small sea urchin *Holopneustes purpurascens* can often be found in groups wrapped in the fronds of kelp. These sea urchins are not usually found in sufficient numbers to kill their host plants.

When *Ecklonia* Forests are destroyed by storms or are eaten by herbivores such as sea urchins, they may regenerate if the disturbance comes in winter or early spring. The Common Kelp reproduces during this time and has the potential to carpet the available space with small plants. If the canopy of the forest is lost at other times of the year, the habitat changes to one in which large brown algae are interspersed in fields of turfing coralline algae in the genera *Amphiroa* and *Corallina*. Also abundant in these areas of TURF HABITAT are smaller brown algae, such as *Zonaria diesingiana* and *Dictyota dichotoma*, and bushy red algae, such as *Delisea pulchra* and *Asparagopsis armata*. Turban Snails (*Turbo torquata*) are most abundant in this habitat in New South Wales.

The shallow margins of reefs are the most familiar and the most studied but in many places, rocky reefs extend down into much deeper water. The sixth habitat recognised in New South Wales is the DEEP REEF HABITAT. In water deeper than 20 m on exposed reefs and in progressively more shallow water in sheltered bays and estuaries, sessile animals cover much of the reef. These often colourful animals are found as sheets covering the reef or as more erect and branched forms. Some of these species will be familiar to divers in all but name, such as the ascidian *Pyura spinifera* which appears clumped around solitary ascidians of another species, *Cnemidocarpa pedata* (the unusual relationship between these species is described in Chapter 16). Black Sea Urchins are found in this habitat but appear to be sedentary and often shelter near larger sponges and soft corals. Not much work has been done in the Deep Reef Habitat.

The recent appearance and rapid spread of an invasive green alga, *Caulerpa scalpelliformis*, in Botany Bay may threaten this Deep Reef Habitat at some locations. With an average fresh weight of over 5 kg/m^2 at 22 m depth, the emergence of this alga has coincided with a marked decline in the cover of sessile animals, particularly sponges that previously dominated this site. Its appearance is perplexing, given that *C. scalpelliformis* is normally found in colder water and has not previously been reported north of Jervis Bay. Further, the large stands found in Botany Bay have not been recorded from any other part of its Australian range.

CONCLUDING REMARKS

The habitats on New South Wales' rocky reefs are defined by the types and numbers of fish, invertebrates and algae found there as well as attributes of the reef. Although the occurrence of some of these habitats is related to depth, for example the Fringe and Deep Reef Habitats, others, such as *Ecklonia* Forests and Barrens Habitats, are found at many depths. Complicating this pattern, exposure to the full force of the waves pushes many of the habitats into deeper water and changes the relative importance of processes that control their character and persistence. Kelp forests and large sessile animals, for example, are found in more shallow water in sheltered areas. Plants on fully exposed shores and in sheltered areas are shaped differently and the forces that limit their numbers and longevity differ at either end of a spectrum of wave exposure (see Chapter 7). Many species, such as some sea urchins and fishes, are influential in determining the habitat in an area. Often these species are herbivores and they exert their influence though grazing on large brown seaweeds. Other species, such as wrasses and limpets, respond to the structure provided and are abundant in some habitats and rare in others.

This brief description of habitats allows little appreciation of the processes that determine where they are found and how predictable they are in space and time. Nor does it provide adequate information on the relationships between habitats and the ways in which fish, invertebrates and algae respond to them. Habitats change, and in order to describe these dynamics we need to understand the processes that create and maintain them. These issues are the stuff of subtidal ecology as a scientific discipline and will occupy much of this book.

String Kelp rises towards the surface at Portsea, Victoria.
Paul Baumann

3 Victoria

Timothy O'Hara, Paul McShane and Mark Norman

From the swirling kelp forests near the coast to the colourful sponge beds of Bass Strait, Victoria's rocky reefs are diverse, bountiful and spectacular. The canyons and arches off Port Campbell, the 'Walls' at the entrance to Port Phillip Bay, and the plunging granite boulders off Wilsons Promontory are dramatic landscapes. Most of Victoria's rocky reefs are on the open coastline between Cape Bridgewater and Gabo Island and typically occur along headlands and promontories, interspersed with long sandy beaches. Small, shallow reefs are found within the sheltered confines of Port Phillip Bay. In this chapter we provide a brief introduction to the diversity of Victorian rocky reefs and outline the major patterns in distribution and abundance of plants and animals across the state.

Storms in the Southern Ocean far to the south of Australia generate enormous waves that are driven into the Victorian coastline from the southwest. Western Victoria receives the full force of these waves. Central and eastern Victoria are relatively sheltered by Tasmania and the islands of western Bass Strait. The average wave energy affecting Portland in the west is five times that occurring at Port Phillip Heads and twelve times that occurring in the lee of Wilsons Promontory. In addition to storm waves, winds across Bass Strait generate weaker waves that can arrive from the southeast or south-west and large swells regularly sweep along the coast from the southwest. These large-scale differences in wave energy are modified at a local scale by the orientation of the coast and differences in the aspect and arrangement of rocks and crevices.

The predominant movement of water across Victoria is related to the tides. Water moves in from either side of Bass Strait on the flood tide and back out again on the ebb tide. The strongest tidal currents occur at each end of the Strait where the islands restrict the flow

Melbourne
Yarra R.
Portland
Port Fairy
Port Phillip Bay
Queenscliff
Western Port
Portsea
CAPE BRIDGEWATER
Port Campbell
MORNINGTON PENINSULA
San Remo
Phillip Is.
CAPE PATERSON
CAPE OTWAY
CAPE PATTON
CAPE LIPTRAP
WILSONS PROMONTORY
Hogan Is.
Lakes Entrance
CAPE CONRAN
Mallacoota
Pt. Hicks
SOUTHERN OCEAN
BASS STRAIT
N

◁
The smooth granite boulders and cliffs of Wilsons Promontory, Victoria.
Ken Hoppen

◁
Map of Victoria showing place names used in the text.

and are weakest in the middle where the east and west flows cancel each other out. Overlaying the tides are the wind-driven currents that flow from west to east, particularly during winter. This brings cool water from the Great Australian Bight into the north of Bass Strait. Eastern Victoria falls under the influence of the East Australian Current which brings warm water southward along the New South Wales coast, eddying to the west as it reaches Bass Strait. This current not only raises the water temperature on east Gippsland reefs but also brings larvae from the north. Consequently, species characteristic of the New South Wales coast also occur in eastern Victoria.

THE COAST

The movement of the Earth's tectonic plates and smaller-scale geological processes have played a role in shaping the reef communities we see today. Sea levels have fallen and risen as the ice ages have come and gone. There are a series of limestone reefs parallel to the east Gippsland coast that appear to represent old shorelines. Bass Strait was completely dry 16 000 years ago when the sea level was over 110 m lower than it is now. Tasmania and Bass Strait then formed the Bassian Peninsula. Melbourne's Yarra River ran around the east side of the low valley that is now Port Phillip Bay before entering the sea south of King Island. Populations of marine animals and plants to the east and west of the Bassian Peninsula were effectively separated, particularly if they were intolerant to the cold and exposed conditions around southern Tasmania. Some of these separated populations have evolved into distinct eastern and western species.

The geology of the Victorian coastline differs significantly from west to east. Basalt is the dominant rock type in the far west from Portland to near Port Fairy and from the Mornington Peninsula across Phillip Island to San Remo. Wilsons Promontory and the rocky headlands of east Gippsland are mostly granite. Sandstone, limestone and calcarenite make up the rest of the coast. These rocks differ in the way they erode and break up — creating reefs of varying structure and complexity. The weathered granite boulders of Wilsons Promontory provide few crevices for the large predators. Basalt on the other hand weathers gracefully, allowing a multitude of creatures to hide under the loose rocks, or within the crevices and cracks that form in the reef. Sometimes reefs are no more than piles of loose cobbles. These reefs tend to be covered in red algae because the longer-lived kelps and sessile invertebrates have little opportunity to establish themselves before they are disturbed by storms.

DIVERSITY ON INSHORE REEFS

Most Victorian reefs are dominated by large brown algae, particularly Crayweed (*Phyllospora comosa*), which forms dense forests that provide much of the character of shallow rocky reefs. Other large brown algae common on exposed reefs include species of *Cystophora*, *Sargassum* and *Seirococcus*. These forests play an important role in sheltering a variety of animals, such as abalone from predators, waves and water currents. The Common Kelp (*Ecklonia radiata*) is also locally abundant but is usually found in slightly more sheltered conditions than Crayweed. The long stipes of Common Kelp hold the fronds off the reef surface and there is often a profusion of understorey species in forests of the Common Kelp and Crayweed.

Beneath the canopy of Crayweed and the Common Kelp, plants are mostly smaller red algae, such as encrusting pink coralline algae, the fan-shaped *Sonderopelta coriacea*, or small bushy red algae in the genera *Plocamium*, *Gigartina* and *Delisea*. Numerous sessile animals compete with the algae for space to attach to the rock, including sponges, anemones, lace corals (bryozoans), feather-like hydroids, tube worms and sea squirts. These occur in bewildering variety; over five hundred species of sponge have been recorded from southeastern Australia and no doubt more remain to be described. Many of these sessile animals are colonial, consisting of thousands of tiny individuals contained within a common structure (see Chapter 16).

The most conspicuous animals on Victorian reefs are the fishes and large invertebrate herbivores, such as abalone, turban snails and sea urchins. Many carnivorous or omnivorous fishes are permanently associated with these rocky reefs, the most abundant being the Blue-throated Wrasse (*Notolabrus tetricus*), Silver and Sea Sweep (*Scorpis lineolata* and *S. aequipinnis*), Old Wives (*Enoplosus armatus*) and diverse leatherjackets (family Monacanthidae). The plankton-feeding Barber Perch (*Caesioperca rasor*) often forms dense schools adjacent to sheer rock faces. Other fishes, such as Senator and Purple Wrasses *(Pictilabrus laticlavius* and *Notolabrus fucicola*), Magpie Perch (*Cheilodactylus nigripes*) and Globe Fish

◁
Crayweed forms dense forests on many Victorian reefs. Phillip Island.
Matt Edmunds

◁
Typical mixed canopy of large brown algae. Wilsons Promontory, Victoria.
Bill Boyle

(*Diodon nicthemerus*), are also abundant. Certain common fishes feed exclusively on the rich algae of these rocky reefs: Herring Cale (*Odax cyanomelas*) and Southern Sea Carp (*Aplodactylus arctidens*) feed on a wide range of seaweeds. The Victorian Scalyfin (*Parma victoriae*) can create patches within the kelp forests by clearing small algal 'farms' in which preferred algae are tended and less nutritious or palatable algae are weeded out (see Chapters 19 and 24).

The most important herbivores are large marine snails, such as abalone (genus *Haliotis*) and Mottled Green Periwinkles (*Turbo undulatus*), and sea urchins. Abalone and Mottled Green Periwinkles eat mainly drift seaweed rather than grazing intact plants. Small abalone can be found in groups in crevices and where drift seaweed accumulates and larger abalone are sometimes found in aggregations out in the open. The most common sea urchin in Victoria is the Purple Sea Urchin (*Heliocidaris erythrogramma*) which may also be coloured green or white. The large Black Sea Urchin (*Centrostephanus rodgersii*) occurs in the far east of the state, from Cape Conran to Gabo Island where it can form large barren areas (locally known as 'white rocks') by clearing and maintaining areas free of foliose algae. Interactions between the grazing activities of sea urchins and the abundance of large marine plants has implications for the composition and diversity of species of subtidal reef communities but are poorly understood for Victorian reefs. Black Sea Urchins compete with Blacklip Abalone (*Haliotis rubra*) in New South Wales (see Chapter 8) and the same processes probably occur in Victoria where the species overlap. Southern Rock Lobsters (*Jasus edwardsii*) may be locally abundant but little is known of the ecological role this species plays.

Many less-visible animals live amongst the dense stands of large brown algae characteristic of Victorian rocky reefs. These include fishes such as the Varied Catshark (*Parascyllium variolatum*), weedfishes (Family Clinidae), seahorses and pipefishes (Family Syngnathidae) and the distinctive Velvetfish (*Aploactisoma milesii*) and Warty Prowfish (*Aetapcus maculatus*). Hermit crabs and large decorator crabs adorned with living algal wigs also go about their business under-cover.

The crevices between and under the rocks also contain a diverse fauna. Large carnivores use these crevices as refuges, such as the Largetooth Beardie (*Lotella rhacina*), Rock Ling (*Genypterus tigerinus*), Red Rock Crab (*Plagusia chabrus*) and the large Maori Octopus (*Octopus maorum*). These species undertake nightly excursions from their shelters to catch fish, crustaceans and shellfish. Crevices in Victorian rocky reefs also contain resident smaller fishes, such as the Tasmanian Blenny (*Parablennius tasmanianus*), Dragonet (*Bovichtus angustifrons*), and the brilliantly coloured Western Blue Devil (*Paraplesiops meleagris*). Other sun-shy animals include the Southern Cardinalfish (*Vincentia conspersa*), assorted scorpionfishes (family Scorpaenidae) and the large orange feather star (*Cenolia trichoptera*).

The obvious animals and plants represent only a fraction of the diversity present on a Victorian reef. There are innumerable tiny creatures that lie hidden among the algae and sessile invertebrates. Many of these animals emerge at night to feed. Diving at night with a torch will illuminate swarms of tiny crustaceans, either swimming around feeding off the plankton or shredding the edges of the kelp fronds. The densely-branched holdfasts which attach the Common and String Kelps (*Macrocystis angustifolia*) to the reefs provide a haven for smaller animals, such as worms, the small green brittle star (*Ophiactis resiliens*), the orange-white sea cucumber (*Pentacta ignava*), tiny bivalve molluscs and crustaceans. Small brightly coloured sea spiders (pycnogonids), sea slugs (nudibranchs) and striped syllid worms can be seen consuming the colonial invertebrates. Unlike many predator–prey interactions in the sea, these relationships can be highly specific, with some sea spiders only found on one type of hydroid or a nudibranch found on only one type of bryozoan.

PATTERNS IN THE DIVERSITY

Despite the geographic differences in topography, sea temperature and history, it is difficult to separate Victorian marine communities into distinct regional groupings. An important reason for this is that there have been insufficient studies of large-scale patterns; the available information does not provide a basis for a comprehensive summary.

Within this constraint, a gradient in the mix of species is evident and several generalisations are possible. Firstly, relatively few species are restricted to eastern or western Victoria, although the ones that have a restricted distribution may be ecologically important. A good example of the latter type is the Black Sea Urchin which causes large

SOUTHERN AUSTRALIAN SEASTARS

BY TIMOTHY O'HARA

Seastars are an important and colourful component of the fauna of subtidal reefs. They include some of the largest predators that live on reefs. A few species can occur in large numbers, and make a significant impact on their prey species. The Eleven-armed Seastar (*Coscinasterias muricata*) can be seen clustered on mussel beds and is considered a pest by mussel farmers. These seastars feed on mussels, attaching their tube feet to either side of the mussel and applying relentless pressure until the two shells are prised apart. Not all seastars eat shellfish. In fact, the majority of common reef species prefer sponges, lace corals (bryozoans) and encrusting colonial sea squirts (ascidians).

Seastars show considerable variation of form and colour. *Echinaster arcystatus* and *Fromia polypora* are the typical five-armed seastar: yellow, orange or red in colour, with a small central disk, long arms and covered in small sharp spines. Large-plated seastars (genus *Nectria*) are also five-armed and reddish orange in colour, but differ in being covered with large round 'plates' on the upper surface. Cushion stars (genus *Patiriella*) and the Velvet Seastar (*Petricia vernicina*) have

changes in the ecology of shallow rocky reefs where it occurs east of Cape Conran. More generally, however, taxa such as the large brown algae that characterise Victorian rocky reefs are broadly distributed across the state.

Secondly, variation among reefs in one region is greater than that among regions. Local factors, such as the difference in wave exposure on either side of a promontory, can be important in determining species composition. Thirdly, species ranges do not coincide. Some east-coast species occur no further west than Mallacoota, others are found only east of Point Hicks, Cape Conran or even Wilsons Promontory. From the other direction, some species are found no further east than Port Phillip Bay, others along the Cape Paterson–Cape Liptrap coastline, and still others further east. These differences in the geographical ranges mean there is a gradual turnover of species from west to east rather than distinct regional boundaries in biological communities on rocky reefs. Lastly, there are few species found only in Victoria. There are many rare species, but they are rare in the sense that they occur in very low numbers throughout their

△
The Eleven-armed Seastar is one of the largest seastars on southern Australian reefs. Victoria. *Paul Baumann*

◁
The Ocellate Seastar showing the large brightly coloured plates on its surface. George III Reef, D'Entrecasteaux Channel, Tasmania. *Simon Talbot*

a relatively large disc and short tapered arms; several species of *Patiriella* have six to eight arms. The biscuit stars (genus *Tosia*) are flattened and pentagonal in shape with enlarged plates along the margin. It is no accident that many seastars are brightly coloured. Their skin contains toxic or bitter substances to deter fish and other predators. The purple and yellow Mosaic Seastar (*Plectaster decanus*) is reputed to cause numbness if carried for any length of time.

Most seastars reproduce by releasing eggs and sperm directly into the water. After fertilisation, the egg develops into a swimming larvae that can spend from a few days to a few months in the ocean before finally settling on suitable habitat to develop into a new seastar. There are many variations on this theme. The Eleven-armed Seastar can divide by fission whilst still a juvenile. A division plane develops across the disc, the arms on either side tug in different directions and the animal is torn in two. Each side grows additional arms and develops into a new seastar. Other species brood their young. The Black and White Seastar (*Smilasterias multipara*) is one of only two seastars worldwide that are known to brood their young in the stomach. The large yolky eggs are laid on the seafloor in September and fertilised. The female then picks up each egg individually with a tube foot and then passes them from foot to foot along the arm into the mouth. The hundreds of swallowed embryos then develop in the stomach over the next few months. The adult presumably fasts during this time as no food is found in the stomach. The tiny juveniles (2 mm across) are released in early November.

Two tiny cushion stars (*Patiriella vivipara* and *P. parvivipara*) are among the rarest marine invertebrates in southern Australia and are restricted to small areas of rocky coastline in southeastern Tasmania and South Australia respectively. Another Tasmanian seastar, *Marginaster littoralis*, has not been seen since the 1970s and is now presumed extinct. Its known habitat along the Derwent Estuary near Hobart has been completely changed by the recent presence of two introduced seastars, the New Zealand Seastar, *Patiriella regularis*, and the North Pacific Seastar (*Asterias amurensis*).

range rather than being limited to a small area.

Although large-scale patterns on Victorian rocky reefs are poorly understood, some patterns on a local scale are better described. As with all coasts, the composition, structure and orientation of the reefs to the prevailing sea conditions have a large influence on plant and animal communities. The most important of these effects are depth, wave exposure and the topographic complexity of the reef.

The first few metres underwater on exposed reefs are dominated by the massive thick fronds of Bull Kelp (*Durvillaea potatorum*). Powered by waves, these fronds scour the rock face, removing many other animals and plants that attempt to colonise. In contrast, sheltered rock platforms are fringed by a diverse range of green and brown algae in the genera *Caulerpa*, *Sargassum* and *Cystophora*. These mixed stands of large brown algae extend down the reef to depths of up to 20 m at some locations, such as the Bunurong coast near Cape Paterson which is protected from storm waves by King Island to the southwest. Sheltered reefs can also support stands of the String Kelp which can grow over 10 m tall.

Below the Bull Kelp, reefs are dominated by large brown algae that compete well in the bright, turbulent waters. The most common species is the Crayweed, which can form dense canopies 1–2 m above the reef and provides much of the character of shallow rocky reefs. The Common Kelp is also locally abundant, particularly in more sheltered conditions.

As light intensity diminishes with depth, animals replace plants as the dominant life form. Sponges, gorgonians, lace corals and ascidians may dominate rock faces. At depths greater than 20 m, such as off Port Phillip Heads and Wilsons Promontory, the fish fauna begins to change with the loss of the kelp-associated wrasses and leatherjackets, and the appearance of fishes such as boarfishes (family Pentacerotidae), Splendid Perch (*Callanthias australis*) and Banded Seaperch (*Hypoplectrodes nigroruber*). Schools of Barber Perch (*Caesioperca rasor*) are replaced by the related Butterfly Perch (*Caesioperca lepidoptera*). In Bass Strait these deep reefs are covered in sponges, bryozoans, large orange fan-shaped gorgonians and exotic-looking animals, such as basket stars (branched brittle stars).

The continental slope (200–4000 m deep) is generally covered in a thick layer of mud, but rocky areas do occur in areas of high water movement such as the 'Cascade' off east Gippsland where in winter the shallow waters of Bass Strait pour down the continental slope. These abyssal reefs are poorly known. Evidence indicates that they are often covered in slow-growing branched corals which shelter mobile animals such as shrimps, sea lilies and brittle stars.

▽ Sea fans, such as *Mopsella zimmeri*, provide colour to many reefs in Victoria. *Paul Baumann*

HUMAN CHANGES

There have been many human-induced changes to subtidal reefs since the explorers Bass and Flinders sailed and rowed into Bass Strait in 1797–98. By this time, sealers were harvesting Australian Fur Seals (*Arctocephalus pusillus*) from colonies at either end of Bass Strait. Although their numbers have increased over the last two centuries, populations of Australian Fur Seals are still much smaller than in pre-European times and it is not known what effect the removal of these once-abundant predators had on the ecology of Victorian reefs. Europeans settled in the early 1800s, and fished the reefs to supply local markets. Fishing pressure has intensified slowly in the early 1900s with the development of diesel engines, and increased demand from local markets. Coastal fisheries boomed after the Second World War, and now reef fishes such as wrasses and morwong, and Southern Rock Lobster, Greenlip Abalone (*Haliotis laevigata*) and Blacklip Abalone are heavily exploited. The extent to which this fishing pressure has caused changes to the ecology of rocky reefs in Victoria is undescribed. Human alterations have also occurred through changes in the catchments of rivers and other land-based activities. Public concern for the conservation of biological resources has brought gradual improvements to the quality of effluent discharges and management of the fisheries. The continuing development of these and other management regimes, including Marine Protected Areas, will help conserve Victoria's rocky reefs.

▽ Bull Kelp exposed at low tide. Mornington Peninsula, Victoria. *Bill Boyle*

Giant Kelp forests provide great complexity to temperate reefs. Tasman Peninsula, Tasmania. *Neville Barrett*

4 Tasmania

Graham Edgar

GEOGRAPHIC PATTERNS

Tasmania's long and rugged coastline has extensive areas of rocky reef. The open coast is broken by many islets and peninsulas that expose the marine plants and animals of Tasmania to a wide range of conditions. The western coast faces the roaring forties and some of the most extreme ocean conditions on Earth. Waves over 19 m from peak to trough, which build up over thousands of kilometres in the southern Indian and Atlantic Oceans, crash onto this coastline. Accordingly, algae and animals living on shallow rocky reefs in western Tasmania require particular adaptations for life in these harsh conditions. By comparison, species found along the northern Tasmanian coast experience milder conditions, with much reduced swells and a smaller seasonal range in water temperature.

Not surprisingly, many contrasts exist in the biological communities associated with reefs between the two regions. In general, the rocky reefs of Bass Strait possess a high diversity of plant and animal species, and relatively low productivity. Reefs on the west coast support fewer species but those present can be extremely abundant. Diving in the dense forests of Bull Kelp (*Durvillaea potatorum*), Giant Kelp (*Macrocystis pyrifera*) and Crayweed (*Phyllospora comosa*) on inshore reefs can be one of the great diving experiences in temperate waters. Bull Kelp and Giant Kelp possess close relatives throughout southern latitudes, including New Zealand, South America, South Africa and the subantarctic islands, and require cold nutrient-rich water to thrive.

Plant and animal communities associated with the eastern and southern Tasmanian coasts are intermediate in form between those in Bass Strait and the western regions. Some common Bass Strait species, such as Horseshoe Leatherjackets (*Meuschenia hippocrepis*) and Victorian Scalyfins (*Parma victoriae*), do not extend around the state's northeastern corner past Cape Naturaliste. Many species, such as the seagrass *Amphibolis antarctica*, Herring

△
A common gorgonian species, *Pteronisis plumacea*, in water 6 m deep at Port Davey; elsewhere in southern Australia this species is not recorded in water shallower than 50 m.
Graham Edgar

◁
Deep reefs support a diverse assemblage of sessile animals. Bicheno, Tasmania.
Matt Edmunds

PORT DAVEY

Port Davey is the only large coastal embayment along the Tasmanian coast between Recherche Bay in the south-east and Macquarie Harbour in the west. Connected to Port Davey by a narrow 12 km long channel is Bathurst Harbour, a shallow 6 km wide basin formed when rising seawater flooded a coastal plain. Waters in these embayments are highly stratified for most of the year, with a brackish surface layer of low-salinity water flowing out to sea above a clear fully-marine layer. The combination of this stratified water with dark tannin-staining of the upper freshwater layer has created an environment unlike any other in southern Australia, allowing glimpses of communities that normally occur below divable depths. The region also possesses unusually low levels of nutrients due to inert rocks in surrounding catchments and a lack of human fertiliser or sewage inputs. Accordingly, the area is of great scientific interest, providing researchers with an opportunity to study particular deep-sea processes, such as the formation of sediments from fragments of bryozoan and other sessile animals (see Chapter 16 for descriptions of these animals).

The major feature contributing to the character of reef communities in Port Davey is a lack of light. Light is rapidly absorbed by the tea-coloured surface water, hence plants cannot obtain sufficient light for photosynthesis at depths greater than 2–3 m through much of the region. On the shallow rocky reefs, seaweeds have been replaced by sessile plankton-feeding animals, such as gorgonians, sponges and ascidians. Other unusual animals, such as basket stars, may be found. Many of these animals normally occur only in depths greater than 30 m elsewhere in southern Australia, so are rarely if ever seen by divers, but in Port Davey they can occur profusely at depths as shallow as 5 m. The fish community also reflects these patterns, with sharks, skates and rays that are common at 50 m depth on the continental shelf replacing the normal mixture of wrasses, leatherjackets and other common reef fishes.

▽
Conocladus australis is the most abundant basket star in temperate Australian waters. Bathurst Channel, Tasmania. *Matt Edmunds*

Cale (*Odax cyanomelas*), Magpie Perch (*Cheilodactylus nigripes*) and the Black Sea Urchin (*Centrostephanus rodgersii*), gradually diminish in abundance down the eastern coast to Maria Island, Tasman Peninsula or Bruny Island. Other species, including those most highly sought after by fishers — Blacklip Abalone (*Haliotis rubra*), Southern Rock Lobsters (*Jasus edwardsii*), Bastard Trumpeter (*Latridopsis forsterii*), Striped Trumpeter (*Latris lineata*) and Warehou (*Seriolella brama*) — are found right around Tasmania, but are typically most abundant in the south. Tasmania also possesses a few species that are found only in the far southern region, particularly the south-eastern embayments and Port Davey in the southwest (see Port Davey Box). These species include Loney's Handfish (*Brachionichthys* sp.), the Many-rayed Threefin (*Forsterygion multiradiatum*) and the Tasmanian Seastar (*Smilasterias tasmaniae*).

The flora and fauna living on Tasmanian reefs thus represents a complex mixture of species derived from several sources. The main components of this mixture are: a Tasmanian element, which is most prevalent in the south, a southern Australian element, which extends across from South Australia into Bass Strait, and an eastern Australian element which extends down from New South Wales with the East Australian Current and diminishes along the Tasmanian east coast and westwards across Bass Strait. The relative importance of each of these components has changed markedly over the past few decades.

The most obvious change in Tasmanian reef communities over the past 50 years has been a huge increase in the prevalence of eastern Australian species at the expense of Tasmanian species. This shift reflects changes in the predominance of the major oceanographic currents that sweep around the island. Most notable among these changes has been the increasing penetration of warm water asso-

▷ Bull Kelp is often the dominant seaweed in the low intertidal–shallow subtidal zone. Bicheno, Tasmania. *Peter Boyle*

▷ Reef with mixed assemblage of brown and red algae, and the ubiquitous Blue-throated Wrasse. George III Reef, D'Entrecasteaux Channel, Tasmania. *Simon Talbot*

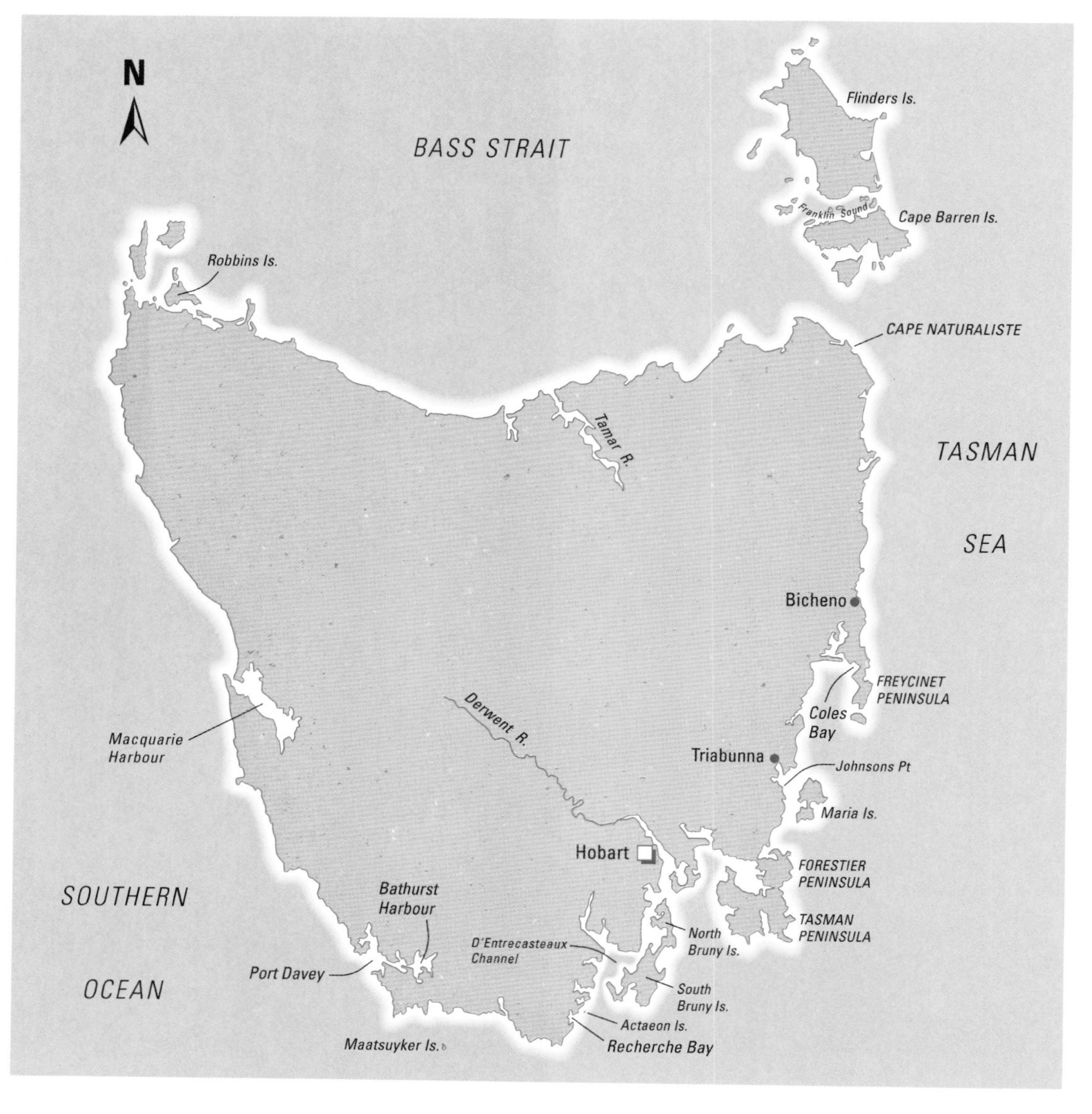

◁ Map of Tasmania showing place names used in the text.

▷ The introduced kelp *Undaria pinnatifida*. Johnsons Point, Tasmania. *Matt Edmunds*

ciated with the East Australian Current along the Tasmanian east coast. The underlying causes of these changes are poorly understood, but are related to the large-scale climate changes that are such an intense focus of atmospheric and oceanographic research.

The increased infiltration of the East Australian Current has brought warmer, clearer water down the east coast, and a declining concentration of dissolved nutrients. This has caused some major changes in the composition of communities on rocky reefs. For example, forests of Giant Kelp that used to be sufficiently abundant to be harvested commercially have greatly diminished in area or have completely disappeared. Animal species that are common in New South Wales but had not been recorded in Tasmania before the 1960s, such as the large Surf Barnacle (*Austromegabalanus nigrescens*) and the Black Sea Urchin, now dominate large patches of reef. Over the same time period, Tasmanian fish and crustaceans, such as Bastard and Striped Trumpeter and Southern Rock Lobsters, have become less common on the east coast, while some typical New South Wales species such as Herring Cale have become more abundant. Interpreting such long-term changes is complicated by intensive fishing in many instances, so for several species it is impossible to identify whether fishing or climate change is primarily responsible for population declines.

The distribution of many species on Tasmanian reefs also reflects the ghosts of past historical events. This is clearly seen with the two species of Australian salmon (*Arripis trutta* and *A. truttacea*). The former species is distributed to the west as far as Western Australia, while the other occurs in the east and north to New South Wales. The ancestor of these two species presumably occurred widely across southern Australia until the last ice age, when the creation of the land bridge connecting Tasmania to the mainland split the population into two. These populations then diverged and evolved into different species. Even though the two species now overlap slightly in their distribution, they have remained distinct.

EFFECTS OF WAVE EXPOSURE AND DEPTH

The pool of species present in a region is primarily regulated by large-scale biogeographic factors, such as water temperature and currents. On a smaller scale the relative abundance of species from this pool on any particular reef is determined by local physical factors, biological interactions and a quotient of chance. The most important of the physical factors are wave exposure and depth but nutrient availability, current flow, rock type, extent of crevices and water turbidity can also greatly affect the distribution of species. Biological factors, such as the supply of larvae, amount of seaweed and distribution of predators and parasites, are also critically important in affecting the abundances of species; nevertheless, these processes operate within the framework set by the physical factors.

The plant communities present on a reef can be predicted to a large extent using information on wave exposure and depth. Sites with a maximal degree of wave exposure, nearly all of which occur in the west and south of the state, almost invariably possess a broad band of Bull Kelp in the low intertidal and shallow subtidal zones. This band varies in depth from 1 m on more sheltered shores to 20 m at the most exposed southerly locations. Bull Kelp can also extend up to 5 m above low tide mark on rocks continuously washed by large rolling swells.

Very few plant or animal species occur in association with Bull Kelp because movement of the heavy fronds abrades the surface, usually leaving only pink coralline algae that coats the rock surface and the encrusting green alga, *Codium dimorphum*. The only common animal is a limpet, *Patelloida victoriana*, and the only conspicuous fish present in the heavy swell is the Purple Wrasse (*Notolabrus fucicola*).

On exposed shores immediately below the Bull Kelp is a band of other large brown seaweeds, with the strap-like kelp *Lessonia corrugata* prevalent in the south and Crayweed in the north. Like Bull Kelp, these seaweeds also normally occur in dense single-species beds, with other species prevented from becoming established by the whiplash action of fronds. A few large invertebrates, including Blacklip Abalone, Cartrut Shells (*Dicathais orbita*) and Warrener (*Turbo undulatus*), maintain a foothold in these forests. A diverse assemblage of small crustaceans, marine snails and other invertebrates live in crevices within the massive holdfasts anchoring the large brown algae to the reef.

Below the bands of Bull Kelp and Crayweed, in depths from about 10–20 m on exposed coasts, is a diverse mixture of large brown algae, mostly in the Order Fucales (eg, the genera *Cystophora*, *Sargassum*, *Acrocarpia*, *Caulocystis*, *Carpoglossum* and *Seirococcus*). Red algae in the genera *Plocamium* and *Phacelocarpus* also occur commonly here, along with searstars, crabs, marine snails and a wide range of fishes, including Bastard Trumpeter, Banded Morwong (*Cheilodactylus spectabilis*) and Blue-throated Wrasse (*Notolabrus tetricus*). In addition to these obvious species, several species of cryptic fish lie concealed amongst the seaweed. Fishes such as the Red Velvetfish (*Gnathanacanthus goetzeei*) and Sharp-nose Weedfish (*Heteroclinus tristis*) are rarely seen by casual observers because of their extremely efficient camouflage.

Providing that waters are sufficiently cool and nutrient-rich, dense Giant Kelp (*Macrocystis pyrifera*) forests are often also found in these intermediate depths. In these forests, plants spiral to the surface from water depths of 5–25 m, using long elastic stipes to form connections from the reef to the surface canopy. Most divers appreciate their similarities with trees and lianas of tropical rain forests, and the way that sunlight shafts in narrow beams to the reef. On the reef, intense competition occurs for light and plants are either adapted for low-light conditions or their growth is retarded until a gap occurs in the canopy, as occasionally happens when a plant breaks away during a storm, dragging others with it. Often, however, the canopy closes over again and many of the plants that respond to more light die off because of the reduction in light.

Giant Kelp fronds at the sea surface buffer most of the effects of wave action, and so provide a sheltered habitat that differs considerably from other reef habitat types in Australia. This habitat is occupied by a wide range of animals, most of them small, cryptic and overlooked by divers. The two major groups of species associated with fronds are the sessile filter-feeding animals attached to leaf surfaces, such as bryozoans and hydroids, and the mobile invertebrates, such as shrimps, copepods and clingfishes. The sessile animals are generally very small and fast growing because they need to complete their life cycle within the time taken for a frond to grow and erode away. Shrimps and other mobile animals are generally larger and move from frond to frond in response to food availability, sometimes reaching very high densities. These animals in turn provide food for fishes such as Bastard Trumpeter, which can form large schools in Giant Kelp forests.

Further down the reef, the Common Kelp (*Ecklonia radiata*) becomes more and more abundant until depths of about 25 m are reached, where it is the predominant plant. Common Kelp has lower light requirements than other large brown algae and survive to depths of 40 m in areas of exposed coast with good light penetration. A diverse mixture of

▷ A school of Bastard Trumpeter off the Tasmanian east coast. *Graham Edgar*

small red and green algae and sessile animals such as sponges are normally also found in these kelp forests. These organisms become increasingly dense with depth to about 35 m on exposed coasts, at which stage the large brown algae rapidly diminish in number. At depths below 40 m, reefs are covered by stunningly colourful and diverse communities of sponges, sea whips, gorgonians, soft corals, ascidians, bryozoans and other sessile animals. Well over 100 species of animal can occur within an area of 10 m^2. These deep Tasmanian reefs rival those anywhere in the world in terms of colour and variety of life, but lie outside normal diving depths so are rarely seen and never studied.

Deep Tasmanian reefs also often support huge schools of fishes, particularly Butterfly Perch *(Caesioperca lepidoptera)*, Barber Perch (*Caesioperca rasor*), Jack Mackerel (*Trachurus declivis*) and other plankton-feeding fishes, with numbers in any area depending largely on the strength of local currents bringing plankton. Reefs located off headlands, or at the entrances of tidal channels, generally possess the largest fish populations, including pelagic fishes such as Barracouta (*Thyrsites atun*), and Australian Salmon that feed on the smaller fishes.

The width of the various habitat zones present on exposed reefs contract rapidly as reefs become more sheltered from wave action. On the most sheltered reefs, Bull Kelp and Crayweed disappear completely. Neptune's Beads (*Hormosira banksii*) and other brown algae, such as *Cystophora torulosa* and *Sargassum* species, predominate in the lower intertidal and shallow subtidal zones. These large brown algae diminish in importance with depth to about 5 m, leaving a few red algae, such as *Thamnoclonium dichotomum* and some sponges. Plants and the reef surfaces are often coated with silt on the deeper reefs because water motion is insufficient to dislodge settled particles, at least until a storm passes. At these sites, seagrass beds often extend offshore from the edge of the reef.

A variety of factors are believed responsible for the relatively low diversity of plants and animals on the most sheltered reefs. The large quantities of silt inhibit the survival of filter-feeding invertebrates due to clogging of feeding mechanisms. Plants in these sheltered habitats also experience great difficulty obtaining adequate nutrients for survival. Although nutrients are generally present in sheltered waters in sufficient quantities for plants, they are rapidly absorbed from the thin layer of water coating the plant surface and this thin boundary layer cannot be rapidly replenished without currents or wave action. Human activities, such as dredging, farming and forestry, also introduce sediments to sheltered rocky reefs, increasing silt loadings and turbidity, and reducing the layer of light penetration for plants.

INTRODUCED SPECIES

One of the clearest signs of change on Tasmanian reefs is the presence of introduced species. Fortunately, rocky reefs have not yet been as severely affected as Tasmanian estuaries which, in the cases of the Tamar and Derwent estuaries, now possess completely different pools of species than were present one hundred years ago. Exotic species, such as the North Pacific Seastar (*Asterias amurensis*), New Zealand Screw Shell (*Maoricolpus roseus*), European Rice Grass (*Spartina anglica*), New Zealand Seastar (*Patiriella regularis*), Pacific Oyster (*Crassostrea gigas*) and Green Crab (*Carcinus*

HANDFISH

BY BARRY BRUCE

Handfish (Family Brachionicthyidae) are curious little fishes that prefer to 'walk' on their fins rather than swim. Their pectoral or side fins are fleshy and the tips resemble human hands (hence their common name). They differ from the related anglerfish (Family Antennariidae) in the form of their first dorsal fin, the location of the gill pore and some aspects of their internal anatomy. Handfish reach a maximum size of between 80–250 mm. Handfish have distinctive lures on the end of their modified first dorsal fin, similar to true anglerfishes and it is possible that these lures play a role in attracting suitable prey. Unlike anglerfish, which are widely distributed in tropical and temperate waters of the Indo-Pacific, handfish are found only in southeast Australia. Handfish can be found on shallow coastal reefs, in estuaries on soft substrates and across the continental shelf to depths of 100 m. Five of the eight currently identified species occur only in Tasmania and Bass Strait. The Red Handfish (*Brachyionichthys politus*) and an undescribed species, Ziebell's Handfish (*Brachyionichthys* sp.) appear to be confined to a few restricted, shallow-reef habitats in southeastern Tasmania. Two other reef-associated 'species', the Waterfall Bay Handfish and Loney's Handfish are possibly colour morphs of Ziebell's Handfish. The Spotted Handfish (*Brachionichthys hirsutus*) is endemic to the lower Derwent Estuary near Hobart and adjoining bays and channels. Spotted Handfish are found on silty-sandy substrates and are only occasionally seen near the reef edge.

Unlike many marine fish that spawn thousands of tiny eggs into the ocean, handfish lay only a small number of large eggs (up to a few hundred) in an interconnected egg mass on the reef. Females wrap the egg mass around structures like stalked ascidians, algae, seagrass or sponge in a process that may take several hours. The female then stays with the fertilised eggs for seven to eight weeks until hatching. Handfish hatch as fully formed juveniles, 6–7 mm long, that drop to the reef and 'walk' away from the egg mass. This reproductive strategy limits the dispersal opportunities for handfish and undoubtedly contributes to their restricted distributions and the possible isolation of colonies. Much of the available information about handfish is limited to the endangered Spotted Handfish, however the reef-associated species from Tasmania are thought to be similar. Red Handfish egg masses and newly hatched juveniles have been observed on green algae (genus *Caulerpa*) and newly hatched juveniles have been seen nearby. Waterfall Bay Handfish egg masses have been observed around sponges. Spotted Handfish spawn in spring and juveniles reach 30–50 mm long after their first year. Handfish sexually mature when they are 2–3 years old and 70–80 mm long, after which growth slows considerably. There are no estimates of how long handfish live. Handfish eat small crustaceans such as amphipods but will accept small live fish in aquaria and have been reported to feed on marine worms.

Handfish have considerable historical significance in Australian ichthyology. The Spotted Handfish was one of the first marine fish collected (in the late 1700s) and described from Australian waters. Yet until recently, very little was known about their biology and ecology and the taxonomy of the group is incomplete. Populations of Spotted Handfish have declined dramatically in recent times and the species now has the distinction of being the first marine fish to be formally listed as 'endangered' under Australia's Commonwealth Endangered Species Protection Act. The species is now the subject of the first federally funded recovery plan for a marine fish in Australia. The recovery plan includes: establishing the cause of their decline, establishing details of their biology and habitat requirements, monitoring the current population and establishing a captive-breeding program with the ultimate aim of reintroducing the species to their former range. The restricted ranges and unique biology of handfish means that all species are potentially at risk of population collapse. All species of handfish are protected in Tasmanian waters.

maenas), now dominate Tasmanian estuaries.

The most serious introduced species on coastal reefs is the Japanese kelp *Undaria pinnatifida*. This species is native to Japan, Korea and parts of China, where it is an important cultivated seaweed. This seaweed probably arrived in ballast water transported by bulk woodchip carriers operating out of Triabunna on the central east coast. Since first noticed in the early 1980s, *Undaria* has spread north to Coles Bay and south to the D'Entrecasteaux Channel, and is currently the most abundant seaweed on many reefs in southeastern Tasmania. *Undaria* has recently also been reported in Port Phillip Bay, Victoria. *Undaria* has proven difficult to eradicate, and is now probably a permanent feature of Australian subtidal communities.

Changes caused by introduced species are possibly far greater than anyone realises — we cannot recognise these changes because no detailed data exists on the natural state of reefs in the distant past. For example, present information is insufficient to judge whether the

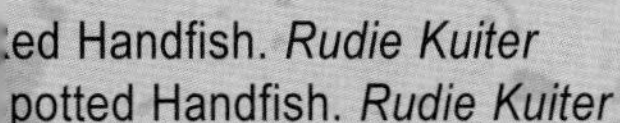

:ed Handfish. *Rudie Kuiter*
potted Handfish. *Rudie Kuiter*

large predatory seastar *Astrostole scabra* is native or introduced. This seastar is abundant on New Zealand reefs and is now common on the eastern, southern and western Tasmanian coasts, but had not been recorded prior to the 1970s. A lack of early records for the species may be due to the seastar not being there, or the species may have been confused with another. Monitoring and control of introduced marine pests are now an important component of Australia's overall management strategy for marine environments.

MANAGEMENT OF TASMANIAN ROCKY REEFS

Managing the diversity of Tasmania's rocky reef resources presents an important challenge. Effective management will require a wide range of options, including Marine Protected Areas (MPAs) which are now recognised as an important tool for the management of coastal resources. As with terrestrial national parks, they can usefully be declared for such reasons as the conservation of diversity, the protection of living resources from over-exploitation, research and public education. In addition, MPAs have the potential to be important sources of larvae, juveniles and adults of fish and invertebrates. Their value as reference areas for identifying the effects of fishing on coastal reefs is also increasingly important. Monitoring reef communities following the declaration of MPAs in Tasmania has revealed huge changes to exploited animal populations once fishing is prohibited. The most obvious effects of fishing on Tasmanian reefs are massive reductions in the populations of exploited fish and rock lobsters. Over the six-year monitoring period at Maria Island, densities of Bastard Trumpeter, the most heavily-targeted fish species on Tasmanian rocky reefs, has increased about one hundred fold within the MPA. Virtually no Bastard Trumpeters were recorded when the MPA was declared, and they remain rare on nearby reefs. Southern Rock Lobsters have increased dramatically in size within all four Tasmanian MPAs that have been monitored. Animals larger than 150 mm shell length are now common within reserves but are very rare outside. This increase in average size, coupled with rising density, has resulted in the total weight of animals within the MPAs increasing by a factor of ten over six years of protection.

Although MPAs have an important role in the management of these resources, they will be most effective in conjunction with more traditional fisheries management methods, such catch limitation and size limits. Further, as seen in the estuaries of Tasmania, adjacent land use and pollution need to be minimised to enhance the prospects of these unique communities surviving in the long term. A wide appreciation of the limits to exploitation of species and an understanding of the diversity and inherent value of rocky reef communities will aid this cause.

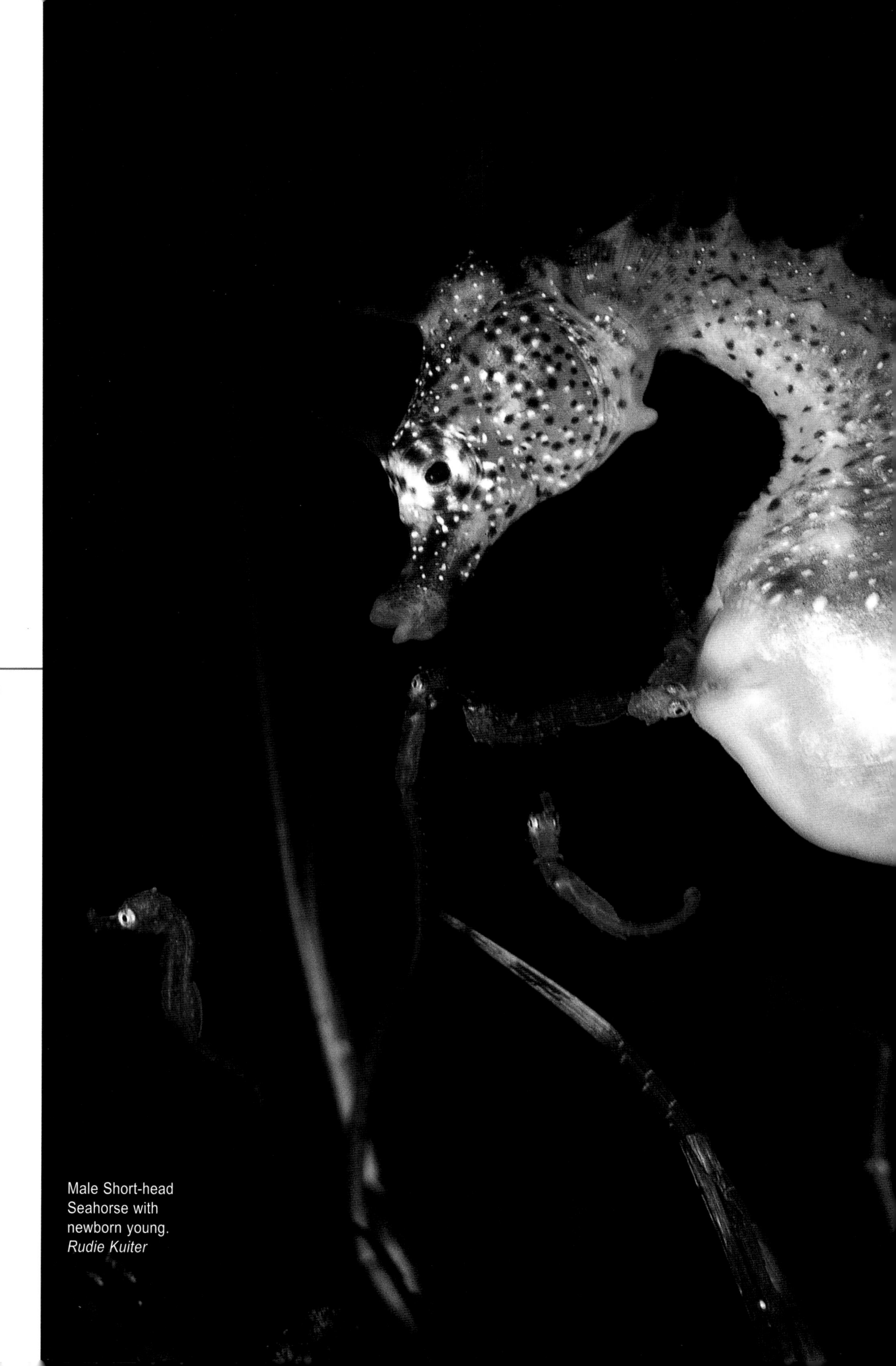

Male Short-head Seahorse with newborn young.
Rudie Kuiter

5

South Australia

Karen Edyvane and Scoresby Shepherd

The marine flora and fauna of South Australia is a mixture of species typical of both the cool temperate regions of Victoria and Tasmania and the warmer waters of southern Western Australia. The combined influences of coastal topography, ocean currents and a long coastline have made this flora and fauna one of the most diverse in Australia. Rocky reefs range from the extensive limestone cave systems of the eastern Great Australian Bight to the spectacular granite headlands of southern Eyre Peninsula and southern Kangaroo Island. In the southeast, cool subantarctic waters provide an ideal habitat for large kelp forests, while strong tidal races at the entrances to the gulfs (Backstairs Passage, Investigator Strait, Thorny Passage and the Troubridge Shoals near Edithburgh) support unique populations of sessile animals.

An important factor in explaining this diversity is the wide range in sea temperatures found in the state. The influence of the warm Leeuwin Current is felt in the Great Australian Bight where water temperatures range between 16°C and 20°C. In Gulf St Vincent and Spencer Gulf, annual temperatures range between 12°C and 25°C. Localised summer upwellings along the southern and western sides of the Eyre Peninsula, and from Robe to Port MacDonnell cause temperature to be lower and to fluctuate less (range 12–18°C). The upwellings in the south-east region, and to a lesser extent the western Eyre Peninsula and western Kangaroo Island, are the most significant upwellings on the southern Australian coastline.

THE COAST

Along the 3700 km coastline of South Australia there is a wide range of coastal landforms and marine habitats. Sandy beaches and rocky shores dominate the coast, comprising about 59 per cent and 33 per cent of the coastline respectively. There are about 150 islands off the South Australian coast, the largest of which is Kangaroo Island. These islands are remnants of the much larger Australian landmass that existed at a time of lowered sea levels during the last ice age 17 000 years ago. When sea levels rose the islands were cut off from the mainland and samples of flora and fauna at that time were isolated with them.

Much of South Australia is semi-arid so there is little coastal runoff from rivers and water clarity is generally in excess of 30 m in the oceanic regions. The Murray River is an obvious exception to this and the plume of fresh water at its mouth is considered by some scientists to be an important barrier to the dispersion of algae and animals along the coast. From the Murray River mouth, the Great Australian Bight, sediment transport is largely caused by ocean swells and storm-generated currents. Within sheltered areas such as the gulfs, redistribution of coastal sediments results in shallow coastal embayments, dominated by tidal saltmarshes, mangroves and seagrass. In the Great Australian Bight and other areas of exposed coast, these reworked sediments form large sand dunes (eg, Canunda National Park).

Rocky reefs in South Australia are either steeply sloping granites or gneissic reefs, or low-profile limestone reefs. Limestone is the dominant rock type in the southeastern region, western Eyre Peninsula and, to a lesser extent, in the gulfs. Granites or gneisses dominate wave-exposed capes and promontories such as southern Eyre Peninsula, southern Kangaroo Island and south-west Yorke Peninsula. Between the granitic headlands, on open coasts and on the west coast of South Australia sandy beaches or limestone cliffs derived from ancient consolidated sand dunes (reaching up to 90 m high along the Nullarbor Cliffs) occur along hundreds of kilometres of coast. In the southeast of South Australia, a series of parallel limestone reefs (or stranded dune ridges), some of them submerged at depths to 70 m, are the only complete record of Pleistocene sea-level changes in the world.

DIVERSITY ON THE OPEN COAST

Subtidal rocky reefs in South Australia are dominated by a diverse array of large canopy-forming brown algae and sessile animals such as sponges, bryozoans, ascidians and hydroids. The algae found on reefs across the state are distinct subsets of species from this pool. East of Robe, the algae are typical of the cold temperate flora of Victoria and Tasmania. Algae such as Crayweed (*Phyllospora comosa*), String Kelp (*Macrocystis angustifolia*) and Bull Kelp (*Durvillaea potatorum*) are the dominant species. Other large brown algae typical of this region include *Seirococcus axillaris, Carpoglossum confluens* and *Acrocarpia paniculata.* West of Robe, the algae are more typical of southern Western Australia. Species in the genera *Sargassum* and *Myriodesma,* as well as *Scytothalia dorycarpa,* are commonly recorded. Superimposed on this transition are the upwelling regions off southern Eyre Peninsula and southern Kangaroo Island which make the transition diffuse by extending the westward limit of cooler-water species. The Common Kelp (*Ecklonia radiata*) is found on reefs along the entire South Australian coast.

Some genera of large brown algae, such as *Cystophora* and *Sargassum,* are diverse and species within these genera show different regional patterns of abundance. Some of the 18 species of *Cystophora* in South Australia are abundant in the east, in cold water (eg, *C. congesta, C. platylobium, C. retroflexa*) and others are found in the warmer waters of the west (eg, *C. brownii, C. gracilis, C. grevillei*). Like trees, these large canopy-forming brown algae often form the major structural component of the reef ecosystem. Under the canopy-forming plants, a diverse assemblage of smaller red, green and brown understorey and turf algae occur, along with sessile and mobile animals.

On the most exposed reefs and in areas of localised, cold-water upwellings such as southwestern Eyre Peninsula and southwestern Yorke Peninsula, there may be more than 100 species of algae on a reef. Diversity is greatest for the red and green algae. In contrast, in areas that are prone to sand scour, such as the head of the Bight and southern Fleurieu Peninsula, there may only be 15–20 species of algae that are commonly found on a reef.

Many of the marine species which inhabit the temperate reefs of Australia have short larval periods and localised dispersal. Consequently, there are many species that are endemic or have small, isolated populations. For instance, the marine flora of Ward Island (western Eyre Peninsula) is very distinctive in the dominance of a number of generally rare or uncommon plants, including red algae in the genera *Glaphrymenia, Kallymenia, Claudea, Solieria* and *Codium.* Tropical algae have also been recorded in other parts of South Australia, particularly at the head of the Bight and in the gulfs. Species such as *Dasycladus densus, Asparagopsis taxiformis, Liagora farinosa* and species of *Laurencia* are

▷ Typical mixture of sessile animals and red algae on a deep reef, South Australia. *Scoresby Shepherd*

▷ The green alga *Caulerpa cactoides* is abundant on reefs that are affected by sand. *Scoresby Shepherd*

restricted to relatively small areas of the coast.

The high species diversity found in South Australia is not restricted to the algae. South Australian waters have the richest assemblage of ascidians or sea squirts recorded in the world, with over 210 described species. Many of these species have been recorded at the offshore islands of the Great Australian Bight region and among the extensive limestone cave systems of western Eyre Peninsula and the southeast. Other marine invertebrates such as nudibranchs or 'sea slugs' are also very well represented in South Australia, with over 500 recorded species. Among the lesser known invertebrate groups, the Investigator Strait and Backstairs Passage regions are home to a wide range of ancient brachiopods or lamp shells, rare free-living corals and bryozoans. Bryozoans or lace corals reach their greatest species diversity in temperate southern Australia, particularly in South Australia, due to the wide continental shelf, where they contribute to about 80 per cent of the total shelf sediments. In this respect, bryozoans are the temperate equivalents of the hermatypic corals of tropical Australia.

The reef fishes of South Australia are generally typical of southern Australian coastal waters. Many of the species recorded in South Australia have also been recorded in southern and southwestern Western Australian waters (see Chapter 1 and 6), and to a lesser extent, in the waters of western Victoria and northwest Tasmania (see Chapters 3 and 4). The most common species in shallow rocky areas are: the Blue-throated Wrasse (*Notolabrus tetricus*), the Silver Drummer (*Kyphosus sydneanus*), Zebra Fish (*Girella zebra*), Dusky Morwong (*Dactylophora nigricans*), Magpie Perch (*Cheilodactylus nigripes*), Herring Cale (*Odax cyanomelas*) and Southern Sea Carp (*Aplodactylus arctidens*). Of the 370 species of marine fishes recorded from South Australia, 77 species are harvested commercially, with 15–20 species contributing the greatest share of the commercial catch.

On exposed coasts sessile animals are not conspicuous and the diver must search in caves or shaded surfaces to find them. In shallow waters (depth 1–3 m) on exposed, shaded cliffs, communities of jointed bryozoans (Family Catenicellidae) form very dense turfs. These filter-feeding animal colonies are flexible like algae and conceal many species of sea spiders or pycnogonids which feed on the bryozoan zooids. Elsewhere on equally exposed limestone coasts with typically undercut platforms, ascidians, rich in species, form communities whose bright and flamboyant colours far exceed those of the better-known tropical corals. In darker places, perhaps with less water movement, like caves, there is a very distinctive fauna. Gorgonian corals in the genera *Melithaea*, *Mopsella* and *Mopsea* and alcyonarians in the genus *Capnella*, together with sponges and hydroids, may cover cave walls, while crinoids like *Comanthus* project their arms from crevices. Crabs are not conspicuous in open-reef habitat because they are a favoured prey of wrasses but they may be seen sometimes in crevices or other places where their predators may be avoided. The commonest reef species are the Red Rock Crab (*Plagusia chabrus*) and the nocturnally active *Nectocarcinus tuberculosus*.

Undoubtedly the most common invertebrate grazers are the abalone (genus *Haliotis*). The five species of abalone found in South Australia live from the intertidal to more than

50 m depth. The most common species are the Blacklip Abalone (*H. rubra*) and the Greenlip Abalone (*H. laevigata*) but two other species may also be found. Roe's Abalone (*H. roei*) lives in sometimes high densities of 10–20/m^2 on exposed and semi-exposed reefs but is usually in crevices to avoid its main predator, the Red Rock Crab. In deeper water to at least 50 m, the much smaller *H. scalaris* may be found in crevices or under boulders to avoid its many predators, crabs and wrasses. Other common grazing animals in algal communities on rocky reefs include sea urchins *Heliocidaris erythrogramma, Phyllacanthus parvispinus* and *Goniocidaris tubaria*, and the gastropods, the Warrener, *Turbo undulatus* and the related species *T. torquatus*. The regular echinoid *Holopneustes purpurascens* is always hidden by the algal fronds to which it clings tenaciously. The Purple Sea Urchin (*Heliocidaris erythrogramma*) and the larger sea urchin *Centrostephanus tenuispinus* occur in the eastern Bight, usually where they can find refuge in a crevice; the latter rarely is found west of the Bight. *C. tenuispinus* has been found in small patches of Barrens Habitat similar to those created by its congener in New South Wales but these are small and relatively rare. In shallow depths (to about 10 m) the Warrener can occur in aggregations at densities to 100/m^2. Its larger congener, *T. torquatus*, is moderately abundant west of Kangaroo Island, while the largest species in the genus *T. jourdani* occurs only in low densities at depths of 20–30 m in the eastern Bight.

Not only do the waters of South Australia contain high levels of diversity and endemism, but they are also becoming increasingly recognised as an area of significance for several species of rare or endangered marine mammals. The waters at the head of the Great Australian Bight represent one of the most significant breeding and calving sites for Southern Right Whales (*Eubalaena australis*) in Australia and there are important breeding colonies of Australian Sea Lions (*Neophoca cinerea*) at The Pages Islands, near Kangaroo Island (see Chapter 26).

DEPTH AND EXPOSURE

Within these broad geographic patterns there are important differences in abundance that are related to depth, exposure and to a lesser extent, rock type. On steeply sloping granite reefs, such as on the southern Eyre, Yorke and Fleurieu Peninsulas, there is a striking three-zone pattern in the abundance of algae with depth. The uppermost zone (0–3 m), which is subject to strong surge conditions, is characterised by dense mats of turfing corallines (genera *Corallina* and *Haliptilon*). Interspersed across these mats are large brown algae, such as species of *Sargassum, Gelidium, Zonaria* and *Caulerpa*, as well as *Cystophora intermedia* and *Myriodesma harveyanum*.

Below this zone there is a middle zone of large canopy-forming brown algae with a sparse understorey of shade-tolerant red algae (eg, *Plocamium, Phacelocarpus, Carpophyllis*) and prostrate algae (eg, *Sonderopelta, Peyssonnelia* sp.). At upper levels of the middle zone, the canopy of large brown algae is formed by the Common Kelp and species of *Acrocarpia* and *Scytothalia*, which are eventually replaced in deeper waters by other large brown algae such as *Cystophora platylobium, Sargassum fallax* and *Myriodesma quercifolium*, among others and depending on location. The lower limit of the middle zone generally occurs around 20–30 m depth, depending on water clarity, but extends to at least 50 m in the clear offshore waters around islands, such as Pearsons Isles in the Great Australian Bight.

Large brown algae below this limit eventually give way to shade-adapted, foliose red algae, which continue in the third and lower zone, which is increasingly dominated by sessile animals until rock is buried by sand. If rock meets the sand at shallower depths than the lower limit of brown algae then the lower, third zone is absent altogether. On the exposed limestone reefs, similar patterns of depth and light-related zonation occur but are made less clear because of sedimentation and sand scour on shallow flat reefs.

This simple scheme is more complicated on limestone reefs where the reef is flatter and more prone to sand scour. On such reefs, such as on the southeast and western Eyre Peninsula, large brown algae in the middle zone are generally less dense on flat reefs in the upper levels (4–10 m) due to sand scour. Exposed reef surfaces are largely colonised by a mixed turfing assemblage of small, sand-tolerant robust red algae (eg, *Gelidium, Laurencia*) and also, species of the green alga *Caulerpa*, which are able to colonise the mobile sediments on reefs. On the southern and western coasts of Kangaroo Island, species of *Caulerpa* form small forests up to 1 m high that compete with the large brown algae.

On exposed limestone reefs in deeper water (10–30 m), the upper zone is dominated by large fucoid algae of genera *Carpoglossum, Acrocarpia* and *Seirococcus*, as well as *Perithalia*

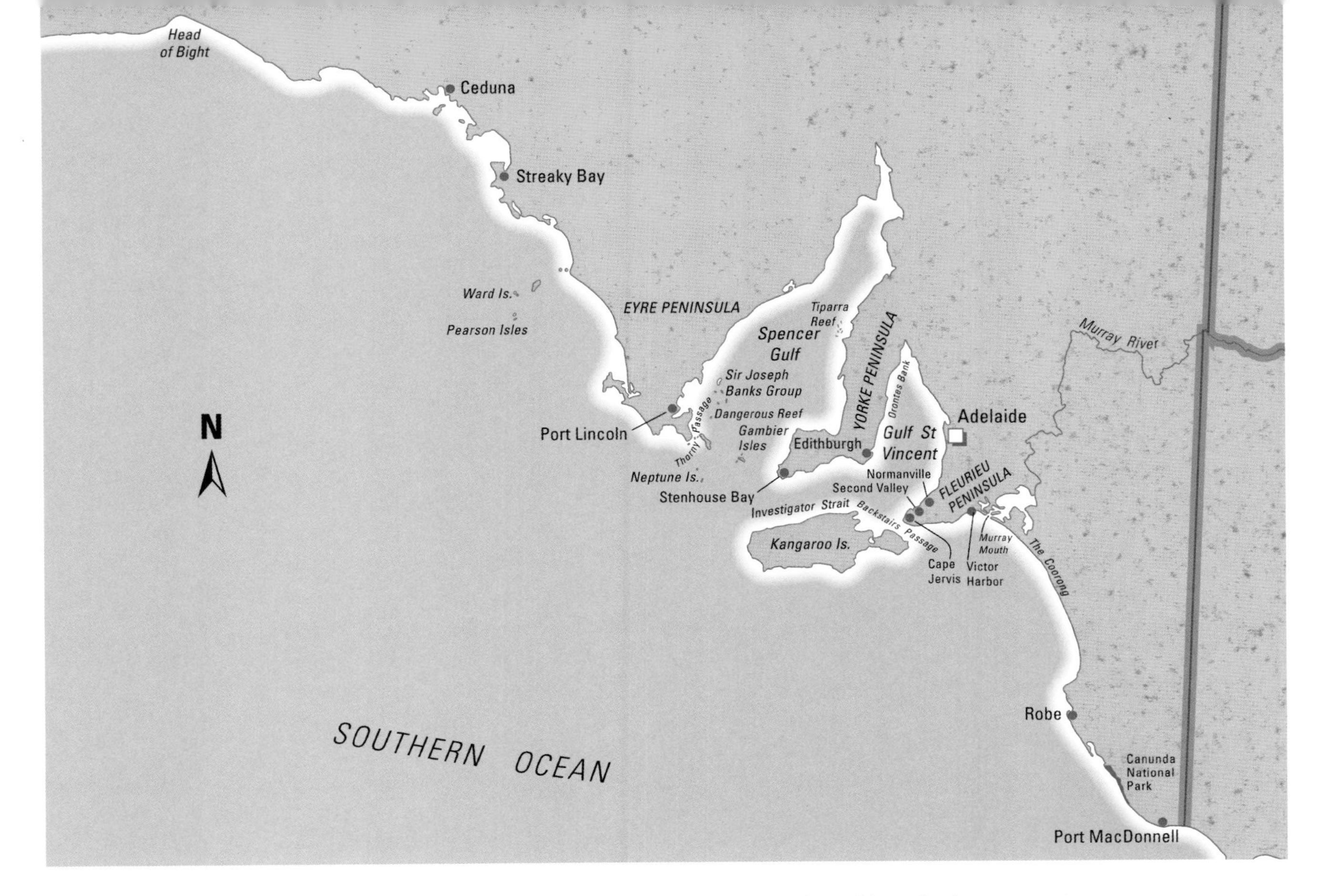

△ Map of South Australia showing place names used in the text.

caudata and *Cystophora platylobium*, which are replaced in deeper waters by Common Kelp and Crayweed. On reefs of low relief (eg, offshore Coorong), there are diverse red algal communities of large fleshy and foliose species, particularly the genera *Osmundaria, Plocamium, Phacelocarpus* and *Melanthalia.* In the southeast, tall forests of String Kelp may be found on reefs at depths of between10 m and 25 m in the lee of nearshore reefs. Below this zone (>30 m), shade-tolerant red algae, such as prostrate algae in the genera *Sonderopelta, Lenormandia* and *Phacelocarpus,* form a conspicuous zone to about 70 m depth, where sponges, soft corals and bryozoans become dominant. A remarkable feature of these waters is the presence of large stands of *Palmaclathrus* at depths of 40–60 m. This is an astonishing green alga that sheds its delicate, membranous fronds annually leaving only a stem attached to the reef. The stem apparently lives for up to 10 years. This alga also occurs in the Great Australian Bight at 25–70 m depth, and even in shallow caves, but not in extensive beds as in southeastern waters.

Very little is known of the deep-water reef communities of South Australia. From the very few detailed studies of deep-water reefs done, it is clear that these animal-dominated communities are also extremely diverse. For instance, the Backstairs Passage region is one of the richest areas for sessile invertebrate fauna in South Australia. Strong current flows in the narrow Passage (up to 250 cm/sec) allow filter-feeding organisms to out-compete marine plants in colonising reefs, and results in one of the few places in South Australia where deep-water reef communities invade shallow depths (ie, up to 15 m). These strong current flows provide optimal conditions for filter-feeders and many organisms, particularly sponges and bryozoans, reach high densities. At greater depths (> 50 m) in the tide race, massive sponges (more than 1 m high and across) and large erect bryozoans, such as the Basket Bryozoan (*Adeona grisea*), dominate the reef. In the sediment drifts, the brachiopod *Magadena cumingi* reaches densities up to $80/m^2$. The deep-water sessile invertebrate fauna of the Passage is of national importance in terms of the diversity of brachiopods and possibly, global importance if records of stalked crinoids are confirmed.

Just as there are species of *Cystophora* and *Sargassum* with differing distributions across the state, species within these genera also differ in their tolerance to wave exposure. Typically, there is an overlapping series of distributions of these species with increasing depth or along coastlines of increasing exposure to the open ocean. On exposed reefs and in shallow water, species such as *Sargassum fallax, S. lacerifolium,* and *Cystophora intermedia, C. racemosa* and *C. grevillei* are most abundant. In deeper and more sheltered water these are replaced by species such as *S. verruculosum, S. sonderi, C. retorta* and *C. subfarcinata.*

SEAHORSES AND PIPEFISHES

BY RUDIE KUITER

There are more than 270 species of seahorses, sea dragons and pipefishes (Family Sygnathidae) worldwide and this number is likely to rise as more is learnt about the group. In southern Australia there are about 34 species, including the most spectacular of all the sygnathids, the Leafy Sea Dragon (*Phycodurus eques*) and Weedy Sea Dragon (*Phyllopteryx taeniolatus*). There are about 45 species of seahorses, ranging in size from just 2 cm in height fully grown to about 35 cm. The tiny Pygmy Seahorse (*Hippocampus bargibanti*) is the smallest and is found only on deep reefs in the tropics. The Big-belly Seahorse (*Hippocampus abdominalis*), is one of the largest, and is common in southern Australian waters. This species lives in shallow coastal habitats and offshore in deep water.

Seahorses and pipefishes are mostly small and slender with stiffened bodies that are armoured with bony plates. They all have a pronounced snout with a small, toothless mouth. The snout is used as a pipette to suck prey from the water. Eyes are generally large and all species feed during the day on small crustaceans, especially shrimps and related animals.

Most species are found on or near reefs with algae, seagrasses or sessile animals and only a few species are known to have strict habitat preferences. There is a misconception that seahorses and pipefishes exclusively associate with seagrass beds. Whilst some are found along the edges of seagrass beds and surveys in such areas may produce large numbers of individuals at some times, very few species associate exclusively with this habitat.

The pipefish fauna of southern Australia is better known than that in many parts of the world. They are more spectacular and generally much larger than tropical species. The Brushtail Pipefish (*Leptoichthys fistularius*) in southern Australian waters is the largest pipefish and grows to be 65 cm long. It lives in protected bays along the edges of seagrass beds but, like most members in the family, it is rarely seen because of excellent camouflage. Australia is very fortunate to have large sections of unspoiled coastline that support the majority of species. Many species live only in small bays in seagrasses. Verco's Pipefish originally described from the Spencer Gulf, South Australia, is now only found in American River lagoon at Kangaroo Island.

Amongst fishes, the sygnathids are unique in that the male incubates offspring on a specially prepared area of the body or in a pouch. Protection of the brood is most developed in male seahorses, which have a pouch with a small opening on the top that is controlled by strong muscles. In other members of the family the degree of protection is less and this relates to the way species live. Those that cling to algae or the reef have the most protection, whilst those swimming and rarely touching the substrate have the least. In the latter, the brood is carried on the outside of the skin and each egg is partly exposed, sitting in a fitting depression that formed when it was deposited. The swimming species are less likely to get attacked by crawling predators such as crabs or worms, but some wrasses and leatherjackets are known to attack broods of sea dragons.

Seahorses breed in early spring and may produce multiple broods in a season. Several species are known to form pairs that stay together for the entire breeding period. They are usually found within sight of each other and greet each other every morning, the male circling the female and both may change colour or the male inflates its pouch, depending on the species. During courtship the female signals her readiness to copulate by lifting her snout upwards and the male responds in the same way. Together they slowly rise and the female places her ovipositor in the pouch opening and the eggs are quickly transferred. As they separate and swim down to the reef, the male twists his body from side to side, positioning the eggs in the pouch so they become attached to the membranes to be fertilised. The largest seahorses produce about 450 eggs in a brood and small species as few as 20. Medium-sized species, such as White's Seahorse (*Hippocampus whitei*) and the Short-head Seahorse (*H. breviceps*), have broods numbering about 100 eggs. Young are of similar size between species and born after 3–4 weeks incubation. The hatchlings of the different species are independent of their parents and have either a pelagic stage, attaching to floating matter, or remain on the reef near their parents. During the incubation time the female develops her next brood and mates with her partner within a few days after he gives birth.

Male sea dragons incubate the brood on the outside of their body with each egg partly embedded in the skin. Prior to mating the area of skin where the male incubates the brood is prepared and conditioned for the eggs. When ready, it becomes slightly swollen, soft and spongy, but soon hardens after the eggs make contact and are pushed on by the female. The eggs are produced as a rolled sheet with several rows of eggs side by side and put on the underside of the male as she moves forward at a slightly faster speed. Although the eggs appear to be adhesive, she may assist, pushing the eggs on by crossing the body of the male with her belly several times. As the sperm comes via the skin of the male, the eggs need to be on the skin to be fertilised. Any eggs falling away will be wasted. Eggs counted on sea dragons were, on average, about 300 to a brood.

Sea dragons usually have only a single brood per season, but under favourable conditions a second brood may be produced. It takes almost two months for a brood of young to develop before hatching over a period of about six days. The hatchlings are quite large when born, ranging from 25–35 mm in length and have a yolk-sac that

▷ Leafy Sea Dragon displaying its remarkable camouflage. *Rudie Kuiter*

▷ Male Weedy Sea Dragon carrying eggs (shown in pink) on his tail. Jervis Bay, New South Wales. *Rudie Kuiter*

supports them for about two days, a time in which the snout grows, enabling them to feed. Juveniles double their length in one week and after 14 weeks they reach almost 15 cm. At that stage they are more adult-like and their growth rate slows markedly. At one year they are about half of their maximum length; it takes two years to reach full size. Adult sea dragons can reach 40–45 cm in length but despite their large size they are easily overlooked. The Leafy Sea Dragon is the most difficult to find in the wild, even where it occurs commonly and in large numbers. Its superb camouflage has given it the reputation of being rare but in many places, from South Australia to south Western Australia, it is locally abundant.

Seahorses have been sought after for centuries as dried specimens for the tourist trade of southern Europe and parts of Asia and for the Chinese medicine trade which consumes millions of seahorses annually. Seahorses and dragons are vulnerable to over-exploitation because they are easily captured and produce small numbers of off-spring compared to most fishes. Special permits are now required to export seahorses, sea dragons and pipefishes from Australia.

SPENCER GULF & GULF ST VINCENT

Subtidal reefs in the gulfs typically comprise large, isolated shoal systems (eg, Tiparra Reef in Spencer Gulf and Orontes Bank in Gulf St Vincent) or shallow inshore reefs. In Spencer Gulf, granite cliffs and submarine outcrops also occur throughout southwestern Spencer Gulf, including the Sir Joseph Banks Group and Gambier Isles. In Gulf St Vincent, granites and gneisses form most of the rocky cliffs that line the southerly and southeasterly margins of the gulf and the Fleurieu and Yorke Peninsulas. Submarine outcrops are rare, other than granite pedestals at the western entrance to Investigator Strait, and a submerged reef southwest of Normanville.

Within the two gulfs, vertical zonation patterns for algal communities are less pronounced than on the open coast, largely because steeply sloping reefs are rare or absent and are replaced by fragmented limestone reefs of low relief. However, reefs in the gulfs are subject to rapid changes in light and water movement regimes and also, greater sedimentation and turbidity with increasing distance from the ocean. Consequently, there is a gradient in the mix of algal species in the gulfs, particularly in the genera *Sargassum* and *Cystophora*. Tiparra Reef extends over about 20 km² in mid-Spencer Gulf and the most abundant species of large brown algae change with depth.

At the shallowest depths, *Cystophora subfarcinata* and *C. moniliformis* are most abundant; these are replaced by *C. brownii* and *C. expansa* at 2–5 m. In water deeper than 6 m, species of *Sargassum* are most abundant, particularly *S. spinuligerum* and *S. decurrens*, and in the deepest water, between 10 m and 12 m, *S. heteromorphum*, *S. lacerifolium* and species of *Caulocystis* cover most of the reef.

The hypersaline and subtropical conditions in the gulfs are unique in temperate Australia and have probably enabled this region to act as a refuge for species commonly recorded further north. Such species may be relicts from earlier periods of warmer-water conditions. The northern Spencer Gulf region is particularly noteworthy for the occurrence of several tropical or subtropical species. This includes species of algae, such as *Acetabularia calyculus*, *Hormophysa triquetra* and *Sargassum decurrens*, and also invertebrates, such as the endemic coelenterates, *Echinogorgia* sp., *Scytalium* sp. and *Telesto multiflora*; and the ascidian, *Sycozoa pedunculata*, which is known only from the upper Spencer Gulf and Investigator Strait region. There are also several species of fish that are endemic or rare outside the gulfs, such as Verco's Pipefish and the Little Pipehorse (*Acentronura australe*). The Tiger Pipefish (*Filicampus tigris*) is known from tropical waters and Spencer Gulf, and Tyron's Pipefish (*Campichthys tryoni*) occurs in southern Queensland and Gulf St Vincent.

▷ Male and female Short-head Seahorse about to mate. Note female with ovipositor and male's open pouch. *Rudie Kuiter*

▽ A tropical gorgonian, *Euplexaura* sp. is a relict tropical species found in the Great Australian Bight and Spencer Gulf. *Scoresby Shepherd*

◁ The green alga *Palmoclathrus stipitatus* is found in deep water where it can be abundant. Plants shed their fronds each year. *Scoresby Shepherd*

Reef habitats are also species-rich in areas of high tidal and current flow. This includes the reefs at the entrances to Gulf St Vincent (ie, Backstairs Passage, Investigator Strait) and Spencer Gulf (ie, Thorny Passage, southwestern Yorke Peninsula) and also, shallow tidal shoal systems. Thorny Passage, Tiparra Reef in Spencer Gulf and the Orontes Bank system in southwestern Gulf St Vincent are some of the largest reef seagrass systems in the gulfs, and are also major areas of commercial fisheries production, particularly for Greenlip Abalone.

In upper Spencer Gulf, algal communities are particularly impoverished. The fucoids *Scaberia* and *Sargassum* sp. are the only major dominants in those communities grazed intensively by the Purple Sea Urchin (*Heliocidaris erythrogramma*).

CONCLUDING REMARKS

In this chapter we have provided a brief introduction to patterns in the diversity of animals and algae on rocky reefs in South Australia. Although these patterns are being increasingly well understood, almost nothing is known of the processes that operate on reefs to determine where species occur and how abundant individual animals and algae are. The exceptions to this are the economically important species, particularly Blacklip and Greenlip Abalones and the Southern Rock Lobster. Very little is known of the linkages among these species, or of the ecological roles they play.

The impacts of human activities on subtidal rocky reefs in South Australia is virtually unknown. Several recent studies however have indicated significant changes to reefs off metropolitan Adelaide. Trampling and harvesting of marine life on intertidal shores has resulted in the loss of marine organisms (ie, fleshy marine algae, molluscs) and a ban on the harvesting of marine life on all rocky shores in South Australia (from high tide to 2 m depth). A recent study has shown a loss of robust larger brown algae such as *Ecklonia radiata, Sargassum* and *Cystophora* from northern reef systems off the metropolitan Adelaide coast and an increase in opportunistic and turf-forming algae, possibly due to increased turbidity and sedimentation from adjacent sewage and stormwater inputs. Of the 25 species of introduced marine pests recorded in South Australia, only the European Fanworm (*Sabella spallanzani*), which has colonised inshore reefs along the metropolitan Adelaide coast, has the potential to impact on rocky reefs on exposed and semi-exposed coasts.

Western Buffalo Bream over reef at Rottnest Island, Western Australia.
Gary Kendrick

6 Western Australia

Gary Kendrick

A COAST IN TWO PARTS

The west coast of Western Australia faces the Indian Ocean and is swept by the warm subtropical waters of the Leeuwin Current. East of Cape Leeuwin, around Albany and into the Great Australian Bight, the south coast is fully exposed to the Southern Ocean and is bathed in much colder waters. The influence of the Leeuwin Current is erratic on this south coast. The flora and fauna on the west and south coasts reflect these differences in water source and temperature and their exposure to wave action. Temperate and tropical influences in the flora and fauna overlap along much of the west coast south of Shark Bay. The species found on reefs on the south coast are more temperate, and are similar to those found in South Australia.

Differences in the flora and fauna between the west and south coasts of temperate Western Australia overlay differences in the geology of reefs and the climate. In this chapter I provide an overview of patterns in ecology of these reefs and of the influence of the Leeuwin Current on the flora and fauna. The ecology of many of the species mentioned in this chapter, such as Western Rock Lobster (*Panulirus cygnus*), Greenlip Abalone (*Haliotis laevigata*), wrasses (Family Labridae), leatherjackets (Family Monacanthidae) and damselfishes (Family Pomacentridae), are described in greater detail in later chapters.

THE WEST COAST

The western coastline, from the Zuytdorp Cliffs south to Cape Leeuwin, is a product of an eastward moving shoreline that developed over geological time. The present coastal strip, offshore islands and subtidal reefs are a series of exposed or submerged ancient sand dunes. The dunes were laid down by wind and represent successive shorelines. They are rich in shells and carbonates from marine organisms. These dunes have been metamorphosed into

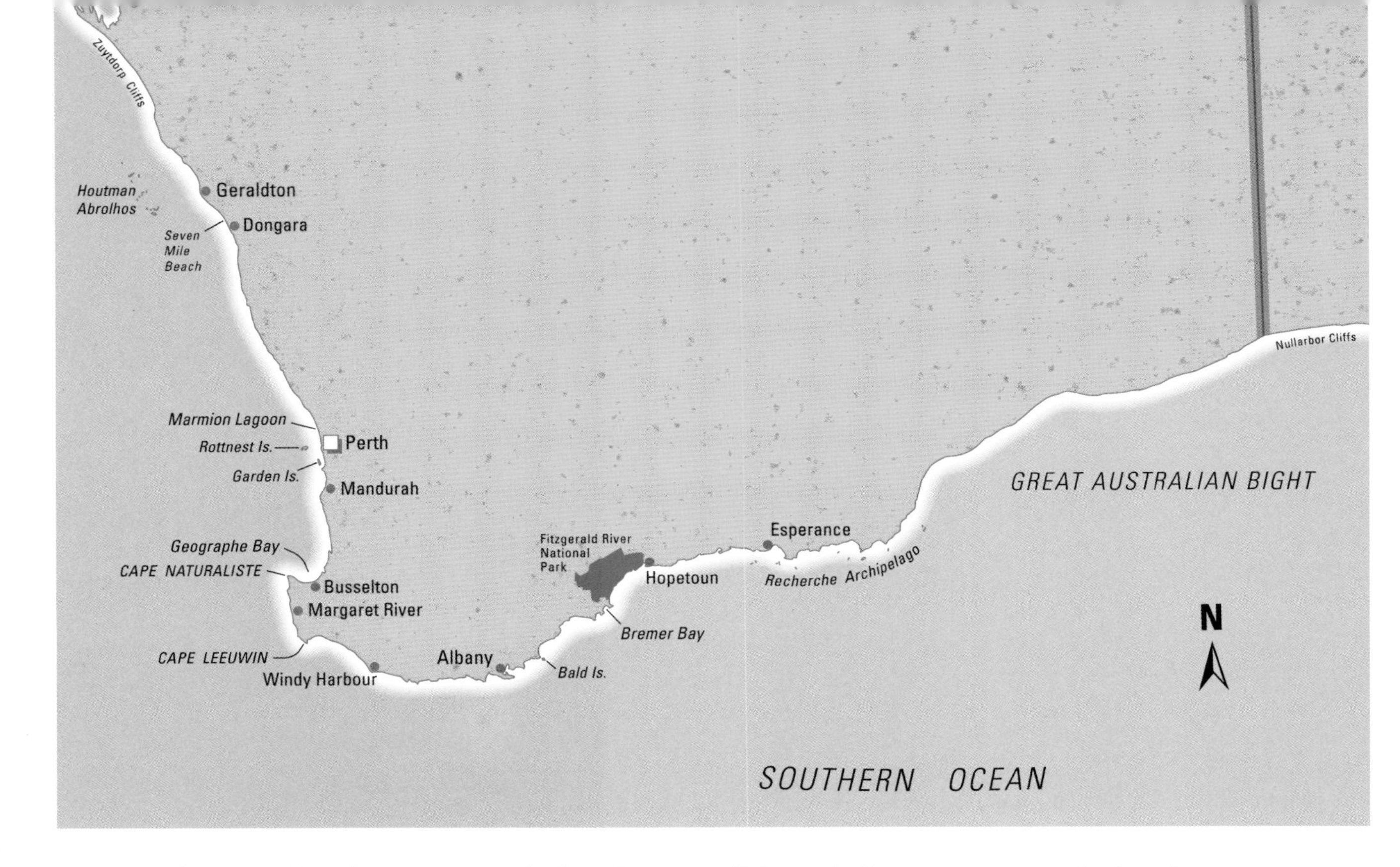

Map of Western Australia showing place names used in the text.

limestone, known as Tamala Limestone, which is found from Shark Bay to the south coast. The westernmost of these dunes formed a series of shallow reefs and islands which run parallel to the coast and are separated from the present shoreline by lagoons and open and closed bays — the largest of which are Geographe Bay, and Warnbro and Cockburn Sounds near Perth. The Houtman Abrolhos Islands, 60–90 km west of Geraldton, are the largest of the offshore islands and support a unique mix of tropical and temperate species (see Box on page 56).

The sea has sculpted the limestone into high-relief reefs with many crevices and fissures rising 2–5 m from the surrounding reefs, or into low-relief pavement, less than 2 m above the surrounding sand. The high relief reefs are dominated by the Common Kelp (*Ecklonia radiata*), which forms a dense canopy over an understorey of encrusting red algae and smaller seaweeds. Where the dense canopy has been removed to expose a patch of reef, many species and forms of seaweeds and sessile animals are found. Resident and pelagic fish species use these reefs for shelter and as sources of food, as do Western Rock Lobsters, abalone and a wide range of other invertebrates.

A rich and colourful mixture of encrusting corals, sea squirts, sponges and encrusting red algae cover the vertical walls, overhangs and caves of these high-relief reefs. These are home to many varied reef fishes. Low-relief pavements are generally covered with large brown algae such as *Sargassum* species. Common Kelp and *Sargassum* occur mixed with various seagrass species — red, green and brown filamentous and foliose seaweeds — and large solitary and colonial sea squirts and sponges. These pavements are subjected to episodes of burial and scouring by sand, producing mosaics of species groups on scales of metres.

Off Perth, a series of offshore islands provide habitat and shelter for seaweed and seagrass beds. Seagrasses develop on the sandy floors of bays sheltered by fringing reefs. The reefs are covered in Common Kelp, *Sargassum* and many species of foliose and turfing red, brown and green algae. The major islands offshore from Perth are, from the south to the north: Penguin, Garden, Carnac, Mewstone, The Stragglers, Rottnest and Little Island. Rottnest Island is 17 km offshore from Perth on the edge of the continental shelf, and is less influenced by coastal processes such as freshwater runoff from rivers and coastal movement of sediment. Rottnest Island is unusual in that it faces north–south whereas the other offshore islands face east–west. The island is bathed in warmer water than inshore islands and reefs. In winter, water temperatures are generally 2–3°C higher at Rottnest Island than on the coast because of the strong southward flow of the Leeuwin Current at that time of year. Rottnest Island has extensive reefs of the coral *Pocillopora damicornis* and one popular snorkelling location on the south side of the island is called Pocillopora Reef. There are also patches of tropical *Acropora* corals on the

▷ Forest of *Sargassum* on limestone reef. Rottnest Island, Western Australia. *Gary Kendrick*

southwest side of the island. In general, the fauna of the island is much more tropical on the western end than in the east.

Within the broad patterns outlined above, fine-scale patterns and the processes that structure communities on rocky reefs in Western Australia are poorly understood. Research has focused on high-profile species, such as abalone and Western Rock Lobsters (see Chapters 8, 9 and 14), and on large brown algae, particularly the Common Kelp and *Sargassum* species. Species and their abundance on these reefs vary with distance from shore. Offshore-Reefs are typically dominated by a dense canopy of the Common Kelp, while those inshore are covered with a diverse patchwork of brown, green and red seaweeds, including Common Kelp, and sessile invertebrates. The difference may be caused by differing exposure to ocean swells, although the complexity and aspect of exposed reefs have an important influence. Sessile animals, such as sponges and solitary sea squirts, are more abundant in areas where the detritus from reefs accumulates.

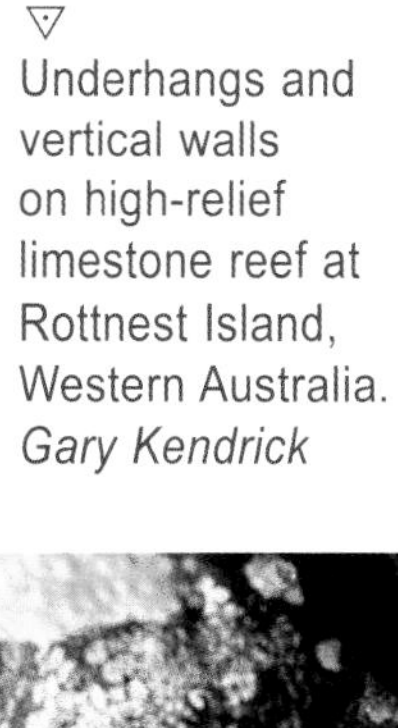

▽ Underhangs and vertical walls on high-relief limestone reef at Rottnest Island, Western Australia. *Gary Kendrick*

SEAWEEDS

Temperate reefs in southwestern Western Australia support diverse assemblages of seaweeds. Of the many seaweeds found, several species of large brown algae are sufficiently large and abundant to form single-species forests. These include the Common Kelp, 18 species of *Sargassum* and more than 20 other species of large brown algae (principally in the genera *Cystophora* and *Scytothalia*). Forests of Common Kelp at Marmion Lagoon account for 95 per cent of the biomass of algae on nearshore reefs. Within these forests, there is a great diversity of smaller algae. For example, 82 species of red, brown and green algae were found associated with forests of the Common Kelp in Marmion Marine Park near Perth.

At a regional scale, there are differences among kelp forests in the types and relative abundance of algae found within the forests. Individual species of seaweeds have restricted ranges across both latitude and the gradient in exposure to ocean swells between offshore, nearshore and onshore reefs. Within kelp forests, storms and other disturbances interact with the ability of algae to colonise gaps in the forest canopy to produce a mosaic of habitats. In a kelp forest, kelp canopies are interspersed with patches of smaller filamentous and foliose algae in clearings. Disturbance and gap creation in kelp forests occur at scales of 1–10 m. Localised dispersal and recruitment across such scales are also common for many seaweeds, including the Common

◁ Stacked sponge (genus *Caulospongia*) on the edge of a reef channel at the Houtman Abrolhos Islands, Western Australia. *Peter Morrison*

Kelp and *Sargassum* species. Dispersal over longer distances can occur, but is more episodic. The combination of local-scale disturbance, dispersal and recruitment results in a mosaic or patchwork of clumps of different seaweeds.

SESSILE AND MOBILE INVERTEBRATES

Sessile animals are major components of subtidal reefs in Western Australia. Yellow, blue and orange blaze from the many species of sponges, ascidians and bryozoans competing for space on these reefs. Under overhangs and in caves, the Sea Squirt (*Herdmania momus*) is a highly visible species on reefs near Perth. They generally feed on detritus derived from kelps and other seaweeds from the reef.

The Western Rock Lobster can be abundant in nooks, crevices and under ledges in the limestone and coral reefs of the west coast, from shallow inshore waters to deeper waters on the edge of the continental shelf. During the day, lobsters are found in the spaces under reefs, but during the night they forage onto reefs and adjacent seagrass beds, travelling hundreds of metres between reefs. Western Rock Lobsters are opportunistic feeders and eat a wide variety of live and dead prey, including molluscs, worms, small crustaceans and seaweeds. This species is described in detail in Chapter 14.

FISHES

Tropical fishes form part of the fish fauna of Western Australia because of the warm Leeuwin Current moving south during autumn and winter. Ninety-eight species (27 per cent) of the fish fauna at Rottnest Island, are tropical; on mainland reefs nearby, only 11 tropical reef fish species are found. Tropical fishes are common in sheltered bays on the southwestern coast of Rottnest Island. Recruitment of small juvenile tropical reef fish occurs sporadically in autumn and winter each year. Some of the hardier species survive to adults. South of Rottnest Island, tropical species have been recorded as far east as the Recherche Archipelago and the Coorong in South Australia.

Schools of Western Buffalo Bream (*Kyphosus cornelii*) and Silver Drummer (*Kyphosus sydneyanus*) are common on all reefs of the west coast of Western Australia, from coastal limestone reefs to the coral reefs of the Houtman Abrolhos Islands. Juveniles are carnivorous, but individuals longer than 0.5 m are strict herbivores and browse on benthic macroalgae, particularly red seaweeds. Silver Drummer feed mainly on brown seaweeds. On shallow subtidal reefs at Rottnest Island, Western Buffalo Bream are territorial in *Sargassum* polygons on shallow inshore platforms, but display their usual schooling behaviour on deeper reefs. These five- to six-sided polygonal hedgerows of *Sargassum* are common seasonally on many shallow subtidal platforms between Rottnest Island and the Houtman Abrolhos. Within a polygon, a single Western Buffalo Bream can usually be found during high tide. These fish are highly territorial, and attack any individual of the same species as well as Silver Drummer, Zebra Fish (*Girella zebra)* and Tarwhine (*Rhabdosargus sarba).* This territorial behaviour is not related to breeding, and may be associated with feeding. The Western Buffalo Bream may also play a role in maintaining the polygons — they remove algae from within them.

▽ Western Blue Devil (*Paraplesiops meleagris*) at Rottnest Island, Western Australia. *Sue Morrison*

DRIFTING WEED

Much of the biomass produced by photosynthesis of seaweeds on reefs breaks off with age and either sinks to the reef or floats. These drifting fragments, which can be as large as a whole plant, accumulate on the sea surface, on beaches and in the swash zone, or in depressions in the reef and seagrass meadows. The Common Kelp drifts along the bottom; the amount of drift kelp may be as much as the total amount of living, attached material at any one time. It drifts for up to 30 km and can be in the water for at least a month before being washed ashore. *Sargassum* thalli float and form rafts that can travel many kilometres and take months to reach shore. These surface-drifting rafts form habitat for juvenile fishes and are at the base of the food chain for pelagic seabirds, such as the Bridled Tern (*Sterna anaethetus*).

Decomposition of this material recycles

HOUTMAN ABROLHOS ISLANDS

The Houtman Abrolhos Islands are made up of three separate limestone platforms situated between 28°S and 29.5°S. These are, from north to south, the Wallabi, Eastern and Pelsaert Island Groups. The Houtman Abrolhos Islands are located on the margin of the continental shelf, 60–90 km west off Geraldton. They are the most southern coral reef complex in the Eastern Indian Ocean and exist because of the influence of the south-flowing warm Leeuwin Current. Winter water temperatures are 4°C warmer at the Houtman Abrolhos than those of adjacent coastal waters near Geraldton.

At the Houtman Abrolhos Islands, kelps grow alongside corals. These kelps predominate on the windward reefs and central lagoons of the Houtman Abrolhos and are the conspicuous seaweeds, along with corals of the genus *Acropora* and large brown algae of the genera *Sargassum* and *Lobophora*. The kelps are smaller, grow more slowly and mortality is similar to or greater than kelps further south. The main reason for this might be the warmer water temperatures: monthly mean temperatures of greater than 23°C occur for the six months from January to June every year. Temperatures higher than 20°C are lethal to most kelps. Tropical components of the flora and fauna (184 species and 42 genera of corals) of the Houtman Abrolhos Islands are well represented, probably because the Leeuwin Current brings low-nutrient, higher-temperature tropical waters to the islands. The Leeuwin Current inhibits nutrient upwelling, which retards growth of large seaweeds like the kelps, thus allowing for coral establishment. The extensive growth of kelp may be evidence of periodic influx of nutrient-rich colder water, from the continental slope.

△ Common Kelp growing alongside coral at the Houtman Abrolhos Islands, Western Australia. *Gary Kendrick*

nutrients into the nutrient-poor waters of coastal Western Australia. Particulate matter from the breakdown of seaweeds also has a direct effect on offshore secondary production. For example, growth rates of offshore benthic suspension (filter) feeders are greatly increased in the presence of organic detritus originating from kelp. Much of the detritus from the breakdown of seaweeds can be exported offshore and is an important food for fishes and benthos in submarine canyons.

Beach-cast seaweeds are also important for the nearshore coastal environment, as well as creating a smelly decomposing mess on coastal beaches. When beach wrack is washed back into the sea during storms, the fragments of seaweed, bacteria and the myriad of small crustacea on the seaweed feed subtidal filter feeders, snails and fish. In Western Australia, the wrack-inhabiting sand flea (*Allorchestes compressa*) is returned to the sea during rough weather and is an important component of the diet of surf-zone feeding fishes such as mullet, whiting, herring and Cobbler (*Cnidoglanis macrocephalus*).

SPECIES INTERACTIONS

Little is known of the strength or importance of interactions among species on West Australian rocky reefs. In other parts of Australia, herbivory, competition and predation among species can be important in determining the relative abundance of algae and animals on rocky reefs. On intertidal platforms near Perth, limpets and chitons graze small algae and structure mid-intertidal communities. This is not the case for all abundant grazers. For example, the Tropical Sea Urchin (*Echinometra matheii*) is abundant on reef terraces on western Rottnest Island, but seemingly has little influence on the abundance of the large brown seaweeds. Similarly, Roe's Abalone (*Haliotis roei*) is common on wave-swept reef terraces, competing

for space with limpets and chitons, yet has little influence on the abundance of filamentous and foliose turfing seaweeds because it feeds almost exclusively on drifting weed.

THE SOUTH COAST

The south coast of Western Australia is fully exposed to the Southern Ocean for much of its 1500 km coastline. The headlands and beaches of the wide and open bays are routinely battered by large oceanic swells generated in the 'roaring forties'. The continental shelf along most of the south coast is narrow — in some places less than 35 km wide. The flora and fauna are almost entirely temperate, although the southward extension of the Leeuwin Current still influences this coast and brings with it some interesting tropical species.

Research on commercial fisheries and surveys of species diversity by naturalists form the basis of our limited knowledge of south-coast reef ecology. The south coast has extreme weather conditions and is inaccessible. Between Albany and Esperance are few towns and few access points to the coast. The biggest towns, Bremer Bay and Hopetoun, flank the Fitzgerald River National Park, which contains 80 km of wilderness coastline. The 1000 km of coastline in the Great Australian Bight, between Esperance and Ceduna, in South Australia, is some of the most isolated country in Australia.

The coastline is made of granite, metamorphosed schists, greenstones and limestone. A series of granite headlands are joined by sandy beaches. On the central south coast near Hopetoun, there are limestone outcrops on the shore, on offshore islands and subtidal reefs. This diversity of substrata for subtidal organisms results in large differences in the dominant seaweeds. Westward towards Bremer Bay on the granite reefs, the Common Kelp is often found with the large brown alga *Scytothalia dorycarpa*. In the centre of the Fitzgerald National Park at Quoin Head, species of *Cystophora* and *Sargassum* co-occur on the schists with the Common Kelp and *Scytothalia*. At Hopetoun and east, the limestones are covered with a mosaic of these species. Such shifts in dominant algae have been associated with decreases in exposure to oceanic swells in South Australia. The continental shelf at Hopetoun broadens, which may afford some protection to the limestone reefs. Alternatively, differences in relative abundance may relate to the change in rock type along the central south coast.

The underwater slopes of the headlands and offshore islands of the south coast are steep and the subtidal landscape has near-vertical walls and fields of spherical granite boulders greater than 10 m in diameter, and sometimes as large as 40 m. Kelps and other large seaweeds dominate this habitat down to 25 m. Deeper vertical walls and overhangs are home to a diverse and colourful array of sponges, ascidians, corals, hydroids and gorgonians. These communities have yet to be intensively studied.

▽ The south coast of Western Australia looking west from the Walpole–Nornalup National Park, west of Albany. *Gary Kendrick*

SUMMARY

The temperate Western Australian coastline consists of the tropically influenced warm-temperate west coast, characterised by offshore lines of limestone reef, and the cold-temperate south coast, characterised by granite reefs. The southward-flowing Leeuwin Current influences the recruitment and survival of tropical and temperate organisms along much of the coastline. Many tropical fish, invertebrates and seaweed species are dispersed southward on the Leeuwin Current. Tropical species are abundant on the Houtman Abrolhos and also at Rottnest Island. Common Kelp is dominant over much of temperate Western Australia, covering sunlit areas of reefs, although on the south coast it generally occurs in patchy mosaics with other large brown seaweeds. Under overhangs and in deeper water, where there is less light, reefs are covered with a display of brightly coloured sponges and ascidians. Many tropical and temperate fish are found sheltering under the kelp canopy or in caves and under overhangs on Western Australia reefs.

Zooanthid, *Zooanthus robustus* colony.
Mick Keough

PART 2
Algae and invertebrates

Giant Kelp forests have many parallels to terrestrial forests and are one of the great diving experiences. Tasman Peninsula, Tasmania.
Kevin Deacon

7 Kelp forests

Peter Steinberg and Gary Kendrick

Kelp forests define the character of temperate subtidal rocky reefs. Like corals on tropical reefs, kelps change reefs by adding structure and complexity to the three-dimensional environment. In this regard they resemble terrestrial forests and, like terrestrial forests, the ecology of kelp communities can be understood in terms of a canopy, understorey and epiphytic species, patches and light gaps. These attributes are moulded by storms, herbivory and the demography of the kelp itself. The unpredictable and interactive nature of these processes make kelp forests some of the most dynamic and fascinating communities in the sea.

The term 'kelp' these days refers to seaweeds in the Order Laminariales, although originally the term (first used in Europe in the 1800s) included a broader range of large brown algae. For the purposes of this chapter, it is useful to consider 'kelp' forests in this broader sense and include fucoid algae (Order Fucales) and Bull Kelp (genus *Durvillaea*). Many of these algae grow to lengths of more than 2 m and also form 'forests'. This more inclusive terminology is appropriate for Australian waters, since the greatest diversity of fucoid algae in the world is found in southern Australia, with over 60 species recorded.

The most important species of large brown algae in temperate Australia include true kelps (species in the Laminariales), such as the Common Kelp (*Ecklonia radiata)*, Giant Kelp (*Macrocystis pyrifera)* and the String Kelp (*M. angustifolia*), *Lessonia corrugata* and the introduced Japanese Kelp (*Undaria pinnatifida*), and fucoid algae such as Crayweed (*Phyllospora comosa*), and less familiar species in the genera *Cystophora* and *Sargassum.* The familiar Bull Kelp is also common in southern Australia and is sufficiently different from laminarian and fucoid algae to be set apart in its own Order, Durvillaeales. The geographical distribution of these species in Australia overlaps substantially but within the general distribution of each species, abundances between locations vary considerably. Like most other seaweeds, kelps are largely restricted to rocky reefs and depths of less than 30 m (occasionally reaching down to 50 m).

Beneath the canopy of kelp forests and elsewhere on subtidal rocky reefs, there is an enormous variety of other seaweeds. There are more than 1000 species of seaweed recorded in Australia and the variety of growth forms and life history stages makes for a bewildering diversity. These seaweeds belong to three main groups or Divisions: the red algae (Division Rhodophyta), the green algae (Chlorophyta), and the brown algae (Phaeophyta). Some genera, such as *Zonaria, Dictyota, Caulerpa, Gracilaria, Corallina* and *Martensia*, are widely distributed throughout Australia, but many other seaweeds have a much more restricted distribution. Thus although there is often a core of shared species, kelp forests in, for example Western Australia and southern New South Wales, are comprised of very different algae.

As with much of the Australian marine flora and fauna, kelp forests are less well studied than those in many other parts of the world. However, in the last 15–20 years, with the rise of experimental and quantitative approaches to the study of kelp forests, and an increase in subtidal research in general, considerable progress has been made.

AUSTRALIAN KELP FORESTS — THE CANOPY

The Common Kelp is one of those species with a wide distribution. It is found on reefs from the Houtman Abrolhos Islands in Western Australia, around southern Australia to Caloundra, Queensland, and also occurs in New Zealand, South Africa and in Oman. The Common Kelp is probably the most abundant seaweed of any kind in Australia. Consequently, it is usually the algal species most familiar to Australian divers. Plants are generally less than 2 m in length and canopies are formed below the sea surface, except in the most shallow water. The species is most abundant on shallow reefs on moderately wave-exposed shores, reaching deeper depths (to 45 m) on coasts exposed to large swells. Most of the ecological studies on Australian kelp forests have focused on the Common Kelp.

Giant Kelp and the closely related String Kelp occur in southeastern Australia, with their centre of distribution in Tasmania. Forests of Giant Kelp are magnificent, with canopies that reach to the sea surface, making them one of the great diving experiences. Giant Kelp occurs mainly in deeper water (8–22 m) whereas String Kelp is found from the intertidal to 10 m depth. Giant Kelps are also common in New Zealand and western North and South America. Giant Kelp, in particular, forms spectacular forests, and in California occurs in extensive offshore beds with plants growing to the surface (at a rate of up to 30 cm/day!) from depths of 20 m or more. Unfortunately, Giant Kelp forests on this scale in Australia are restricted to Tasmania. The final species of true kelp found in Australia, *Lessonia corrugata*, is also restricted to (southern) Tasmania.

Bull Kelp is found only in New Zealand, southern South America, some subantarctic islands and Australia, where it is restricted to southeastern Australia and Tasmania. Bull Kelp contains large amounts of alginate (a gelling agent or emulsifier used in food products and cosmetics) which make the fronds elastic and allows this alga to survive in the most exposed intertidal and shallow subtidal habitats.

Algae in the genera *Sargassum* and *Cystophora* are diverse (over 18, and 23, species respectively in temperate Australia). These algae will be familiar to divers in everything but name; plants are typically large, leafy and olive coloured and are found throughout temperate Australia, generally on sheltered to semi-exposed reefs. Individual plants have more structural complexity (stems, blades, floats and the like) than other Australian canopy-forming algae (with the exception perhaps of Giant Kelp). Some stems on these plants have become differentiated into reproductive structures and are lost at the end of the reproductive season. In many species of *Sargassum* there are annual cycles in the size of plants and some over-winter as small rosettes, subsequently growing to be longer than 1 m in summer. *Sargassum* species are important canopy-forming algae on subtidal reefs in both tropical and temperate Australia, and in some areas can be more abundant than the Common Kelp. At Rottnest Island, Western Australia, *Sargassum* covers 20 per cent of the substratum to 10 m depth, and accounts for more than 20 per cent of both the annual production and biomass of algae. Crayweed, another fucoid, is common on exposed coastlines in shallow waters (0–12 m depth) in Victoria and southern New South Wales but is also found in deeper waters to 18 m depths in Tasmania. Individual plants can exceed 4 m in length and form dense canopies.

▷ *Cystophora monilifera* at Stream Beach, Western Australia. *Gary Kendrick*

◁
Crayweed at Phillip Island, Victoria.
Matt Edmunds

▷
A mixed canopy of large brown algae, typical of many reefs in southern Australia. Wilsons Promontory, Victoria.
Bill Boyle

◁
A canopy of Common Kelp, Tasman Peninsula, Tasmania.
Peter Boyle

▷
Delisea pulchra, a red alga common in southeastern Australia which produces natural antifouling compounds. Sydney, New South Wales.
Rocky de Nys

Given the size of kelp plants and their often high densities, it is not surprising that these species can have a strong effect on their local physical environment. However, because of the difference in size, weight and shape of these seaweeds their degree of influence varies. Giant Kelps can occupy the entire water column (to depths of 20 m or more) from the rocky reef to the surface of the water, where their fronds can spread to cover nearly 100 per cent of the surface. They substantially modify the influence of currents and ocean swells, and reduce the amount of light reaching understorey plants. As much as 90 per cent of the light present at the surface can be lost under these canopies. In addition to loss of light, Bull Kelp can have a large impact on understorey animals and algae by abrading the reef surface. The fronds of this alga are large and heavy — individual plants can weigh over 5 kg and the biomass of stands of Bull Kelp can exceed 100 kg/m^2. As fronds are thrashed about in the surge and waves they can inhibit or kill other reef organisms by whiplashing the reef. Crayweed has a similar but less pronounced effect on understorey species.

The Common Kelp is smaller than these other species, but occurs in very high densities (in excess of 20 kelp/m^2 are common). This kelp modifies a much smaller vertical amount of the water column, but has a substantial effect on the amount of light which penetrates the dense canopy. Species of *Sargassum* and some *Cystophora* can also strongly shade the bottom, but often have air vesicles on their fronds and so tend not to drape and scour the reef.

UNDER THE CANOPY

Under the canopy of a kelp forest there is a diverse assemblage of red, brown and green algae. They occur in layers, with the bottom-most layer consisting of encrusting algae (eg, genera *Hildenbrandia* and *Corallina*) which cover the rock and compete with sessile animals, such as sea squirts in genera *Pyura, Botrylloides* and *Didemnum* and sponges in the genera *Chondrilla, Darwinella* and others — see Chapter 16. The next layer, growing up off the bottom to a height of a few to tens of centimetres, consists of a diverse understorey of foliose, filamentous and fleshy algae. There may be as many as 100 species of such seaweeds on a single reef. Common understorey algae include plants in the genera *Zonaria, Lobophora,* and *Martensia australis* in New South Wales, *Plocamium* and *Corallina* in Victoria and South Australia, *Pterocladia lucida, Amphiroa anceps* and *Dictymenia sonderi* in Western Australia, and *Plocamium, Callophyllis* and *Sonderopelta* in Tasmania. These algae often have patchy distributions, and quite different suites of understorey species can occur metres apart on a single reef.

The abundance of understorey algae is heavily influenced by the density of the canopy and the species forming the canopy. There is generally more algae and greater diversity of species in the gaps in canopies and some species appear to specialise in occupying disturbed patches inside kelp forests. Other species may be found under a wide range of kelp densities and can be considered true

ALGAL LIFE HISTORIES

WHY SEAWEEDS ARE NOT LIKE YOU AND I

Seaweeds, along with many other marine organisms, have more than one free living, multicellular phase to their life cycle. In fact, seaweeds pretty much have every variety of sex and reproduction known. There are male plants, female plants, plants that are both male and female, and plants which are neither male nor female (asexual). Moreover, as well as reproducing like us via the production of single-celled gametes (eggs and sperm) which then fuse to form zygotes, many algae also clone themselves. They do this either by producing asexual single-celled spores, or by budding off multicellular fragments of themselves that persist as individual plants, like strawberries. This diversity of life histories is probably one reason why the algae have been so successful in the marine world; over evolutionary time they have responded to different environments by radically changing their reproductive biology and mode of life.

True kelps (species in the Order Laminariales) have one type of life cycle, and fucoids another. Laminarian algae exhibit 'alternation of generations', in which a diploid (having two copies of each chromosome, like humans) sporophyte phase alternates with a haploid (one copy of each chromosome, like bacteria) gametophyte phase. The sporophyte is the large plant we think of as 'kelp'. The gametophytes are microscopic filaments, known mostly from laboratory studies. They are presumed to occur in cracks and crevices on the bottom, but are rarely if ever observed in the field because of their size, and because their simple morphology makes them easy to confuse with other kinds of small algae. Although unseen, gametophytes are crucial to the ecology of kelp forests. For example, they can function like seed banks of terrestrial trees by providing a source for regeneration of kelp forests even when there are no visible kelp plants (sporophytes)

△
Under a Common Kelp canopy. Rottnest Island, Western Australia. *Sue Morrison*

'understorey' species. Such species include encrusting coralline algae, which give the rocks under the kelp canopy its characteristic pink to purple hue. These algae grow as crusts on the reef, removing calcium carbonate from the water and sequestering it in the form of calcite into their bodies. They are thus 'rocks that photosynthesise'. Foliose red algae, such as species in the genera *Pterocladia, Callophycus* and turfing coralline algae (*Amphiroa*), also occur in low densities as understorey species under the kelp canopy.

KELP FORESTS AS DYNAMIC SYSTEMS

Throughout their lives kelp plants are subjected to the destructive forces of storms and waves; they are eaten by herbivores, colonised by epiphytes and infected by disease. The consequence of all these factors is that kelp forests are changing, dynamic systems, and an understanding of the factors which bring about these changes is essential to an understanding of their ecology. Before we consider the external factors which cause changes in kelp beds, however, it is useful to first understand the basic demography of kelps. Of particular importance is an understanding of their complex reproductive biology (see Box below).

DEMOGRAPHY OF KELP

The details of algal sex, or whether a species has one or two distinct phases in their life history, are the qualitative features of algal life histories. Equally important in understanding the ecology of kelp forests are the numbers of new individuals arriving in a forest, and the details of their subsequent survival, growth, reproduction and death. Such quantitative aspects of kelp populations are termed demography, and it is factors which affect the demography of kelp that largely determine the patterns we see in kelp forests.

Kelp have the potential to be reproductive machines. Even relatively small kelps such as the Common Kelp can produce millions of zoospores each time they reproduce. The numbers of spores produced is usually positively

around. The presence of this microscopic phase also has considerable implications for the introduction of new species of kelp (or other algae) into Australia, such as the recently introduced Japanese Kelp. A 4 m long sporophyte of this species on the hull of an overseas freighter is pretty obvious; a gametophyte less than a millimetre long is not.

Of all the algae, fucoids and Bull Kelp have life histories most similar to ours. In these algae there is no multicellular gametophyte. Rather, sporophytes produce eggs and sperm which fuse and form a zygote, which usually remains attached to the parent plant during initial development before settling and attaching to the reef. This germling then develops and grows into a new sporophyte. The life histories of this group can still be complicated, however, because some fucoids have separate male and female plants, while in other species single plants can contain separate male and female reproductive fronds on the same plant (as in *Sargassum* and *Durvillaea*).

Beyond the kelps and fucoids, the diversity of reproduction and life histories in algae reflect the diversity of the plants themselves. In many red algae, rather than sperm and eggs forming a zygote which then becomes a new free-living sporophyte, they first form a small but multi-celled 'carposporophyte' which is held on the female gametophyte. The carposporophyte subsequently divides to produce spores, which then drop to the reef and grow into new sporophytes. In some algae (eg, *Delisea* and *Dictyota*) the gametophytes and sporophytes look identical, while in others they look completely different, as in kelps. Some red and brown algae alternate between erect blade-like or bushy phases and crustose phases. The frequency and timing of sexual reproduction versus asexual reproduction also varies enormously, and can be altered by changes in environmental cues such as light or temperature. Finally, sexual reproduction has never even been observed in some algae, which appear to endlessly clone themselves by recycling asexually produced plants.

△ *Holopneustes purpurascens* in a Common Kelp frond. Forestier Peninsula, Tasmania. *Matt Edmunds*

related to the size of the kelp plant. Ultimately, however, because of the nature of the kelp life history (see Box on page 66), the appearance of new kelps (sporophytes) into the population depends upon sperm finding and fusing with eggs, which in kelps are held on the female gametophyte. The density of spores released into the water column will influence the numbers of microscopic female gametophytes on the bottom, which in turn will influence the numbers of new kelps which recruit into the forest. Although remarkably prolific in producing zoospores, evidence from Australian kelp forests and elsewhere indicates that kelps are mostly not that effective in colonising new areas of the bottom that are more than a few metres from the existing forest. This suggests that even though millions of zoospores may be released from a forest, they are either rapidly diluted in the water column, or largely unsuccessful at surviving the gametophyte stage. Thus once outside the forest, the probability of eggs and sperm fusing to form new sporophytes is generally too low to create a new patch of forest. This implies that, aside from the impact of other ecological processes such as herbivory, the borders of kelp forests should not greatly expand and contract over short time periods. In support of this prediction, in some areas of New South Wales where long-term records have been kept, the general size and shape of many Common Kelp forests have indeed proved remarkably stable over the past 10–15 years.

To add to the uncertain future of a small kelp plant, survival of new sporophytes underneath a canopy is also often poor. Densities of juvenile plants under a canopy rarely exceed 10 per cent of adult densities, primarily because of the shading effects of the canopy. Thus new kelp plants are between a rock and a dark place: outside the forest on the relatively open substratum densities of gametophytes may be too low for recruitment and there is often a greater chance of being eaten by urchins and gastropods and inside the forest there is not enough light. Most successful colonisation therefore occurs in light gaps within the forest itself. However, even the presence of such gaps does not ensure successful establishment of new recruits. In New South Wales at least, the timing of the formation of these gaps is important. The reproductive cycle of Common Kelp in New South Wales is such that production of zoospores peaks in autumn; they then take several months to develop into gametophytes, resulting in a peak of new sporophytes in winter and early spring. If gaps in the canopy are formed in winter (often due to storms), they are rapidly filled by Common Kelp and the canopy is re-established. However, if gaps are formed at other times of the year, they are colonised by turfing algae and understorey species such as the smaller brown algae (eg, genera *Zonaria* and *Lobophora*). These then monopolise the available space, inhibiting the recruitment of juvenile kelp for several years. The timing of the formation of gaps in the canopy may be less critical in Victoria and Western Australia, where zoospore release occurs for most of the year.

Other than these results for Common Kelp, there have been few studies of the demography of other Australian kelps or fucoids. The evidence that does exist suggests that the presence of nearby adults is necessary for successful recruitment (as opposed to recruits coming from longer distances away) in many species, but that established canopies or adults also inhibit recruitment or survival of juveniles. For example, harvesting of adult Bull Kelp in New Zealand results in a reduction in its intertidal range, but the remaining holdfasts of adult plants also still inhibit the recruitment of new plants. *Sargassum* species also appear to disperse locally, with propagules settling only 1–10 m away from parent plants. As with other species, however, established adults negatively impact on the survival of recruits through shading or, for some species, whiplash. Persistence of existing stands of *Sargassum* may in fact be largely due to vegetative (asexual) regeneration from fragments of holdfasts left on the reef.

Growth of kelp can be remarkably rapid. These algae are among the most productive large plants in the world and can produce new tissue at a rate of more than 3 kg dry weight /m^2/day. This is comparable to the productivity of tropical rain forests or of many agricultural crops. Growth of these algae is also usually highly seasonal, and experienced divers will recognise the dramatic increase in the biomass of kelp that occurs in late winter or early spring. Growth then declines over summer and early autumn, and kelps are typically shrinking by late autumn, when much of the frond tissue senesces and falls off the plant. This pattern of growth and senescence is observed for many of the canopy-forming species, although in the case of *Sargassum* it is also due to the production and subsequent sloughing of specialised reproductive fronds.

Persistence and longevity of large brown algae in Australia varies substantially, but individuals rarely live to their physiological life span. Most are killed by herbivores or storms or even adverse temperatures before reaching this theoretical age limit. For example, Giant Kelp is generally found in cold water and does not prosper in warmer conditions.

Giant Kelp plants can live seven years, although most only live 3–4 years. The Common Kelp can also live for seven years, although again few individuals achieve this age. In Western Australia, between 56 and 92 per cent of mature kelps die within 12 months, and most thalli do not survive more than two years. A similar general pattern is true in New South Wales. Bull Kelp can also live for at least eight years, as can species of *Sargassum*, although many populations appear dominated by individuals which live for less than three years.

STORMS, LIGHT GAPS & HERBIVORES

Storms play a fundamental role in the ecology of kelp forests, both in Australia and worldwide. In Australia, such effects are best known for Common Kelp in New South Wales where storms create gaps in the canopy by tearing plants from the reef. In rare instances entire forests can be removed, as was the case for a small kelp forest at Cape Banks near Sydney in the winter storms of 1993. Once these gaps are created, they are colonised by a wide variety of species depending on when and where the disturbance has occurred. The impact of the timing and severity of the storms and subsequent creation of patches without kelp has been discussed earlier in this chapter.

OIL TANKERS, ANTIBIOTICS & CHEMICAL WARFARE IN THE SEA

In an era of economic rationalisation, why care about algae (other than the fact that they create the dominant habitat and food source for many reef organisms)? One excellent reason is that kelp and other algae produce a variety of chemicals that can be used in applications ranging from antifouling paints for boats to ice cream. For example, agar, carageenan and alginates, which are cell-wall constituents of red or brown algae, are used as gelling and emulsifying agents in many applications including ice cream and bacterial culture media. Bull Kelp is harvested commercially on King Island in Bass Strait for its alginates.

Other potential uses of chemicals produced by the algae lead more directly from the ecology of the plants. As described above, seaweeds get fouled by a diversity of epibiota. This same problem is faced by structures humans put in the sea (marine bacteria also colonise these structures, and fouling by bacteria is a problem not just for boats but also for structures ranging from water-treatment plants to artificial heart valves). Fouling is in fact a major problem for all marine industries, with worldwide annual costs to the shipping industry alone in the order of A$7 billion/year due to the added fuel use and reductions in efficiency of operation. By far the most common solution to preventing fouling is the application of toxic, heavy metal-based (primarily tin or copper) antifouling paints to submerged surfaces which slowly leach heavy metals into the environment and kill settling fouling organisms. Unfortunately, such paints also have a variety of wider environmental effects on non-target organisms and are a major source of marine pollution.

Some chemicals produced by seaweeds offer a potential solution to the problem of this widespread use of toxic antifouling paints. Just as fouling of boat hulls leads to slower boats; fouling of seaweeds, if severe enough, can kill these algae. Kelps and other seaweeds have evolved defences which inhibit the settlement and growth of organisms on their surfaces. Such defences include the production of surface slime, occasional sloughing of surface layers of cells, and the production of inhibitory chemicals which function as 'natural antifoulants'. These natural antifoulants are often biodegradable, and also can function as settlement inhibitors rather than toxins (from an evolutionary perspective, the seaweed must avoid poisoning itself). A number of research laboratories throughout the world are currently investigating the use of natural antifoulants from algae or sessile invertebrates as alternatives for the active ingredients in current antifouling technologies. One promising source of such natural antifoulants is the red alga *Delisea pulchra* which is found in southern Australia.

More generally, seaweeds live in a chemical soup of small organic molecules, to which they are active contributors. As well as chemically defending themselves against fouling, algae must also defend themselves against herbivores and bacterial pathogens. Thus many of the unusual compounds produced by algae may also act to deter attack by herbivores — sea urchins, fishes, snails and crustaceans — or as natural antibiotics by inhibiting colonisation and subsequent attack by bacteria. An understanding of the role these chemicals play in inhibiting bacteria in natural systems may lead to the development of new antibacterial compounds for medical and industrial use. In an era when bacteria are developing resistance to antibiotics at an increasing rate, novel sources of antibacterial agents are of fundamental importance.

Herbivores have a range of effects on algal communities. At their most extreme, herbivores may largely determine, or substantially change, the overall boundaries of kelp forests or other algal communities, as in the case of the Black Sea Urchin (*Centrostephanus rodgersii*) or the Purple Sea Urchin (*Heliocidaris erythrogramma*) in New South Wales. High densities of the Black Sea Urchin can denude large areas of kelp forests, resulting in areas dominated by crustose coralline and turfing algae. The extent and nature of grazing by the Black Sea Urchin is described in Chapters 2 and 15.

The Purple Sea Urchin can also dramatically affect the abundance of algae on reefs, changing them from communities largely dominated by foliose algae to ones dominated by encrusting algae, invertebrates and tube worms. Such an event occurred near La Perouse in New South Wales during 1998, when a moving 'urchin front' of sea urchins removed a succession of foliose algae off the bottom, leaving only coralline crust and bare rock. Interestingly, the sequence of removal of algae by the urchins was quite consistent across different parts of the front, with brown algae such as *Sargassum* and *Zonaria* consumed first, then turfing coralline algae such as *Corallina* and *Amphiroa*, followed by the removal of delicate foliose red algae *Delisea pulchra*, until only coralline crusts remained. Although such sea urchin fronts have been described as being analogous to the outbreaks of Crown of Thorns Starfish (*Acanthaster planci*) on coral reefs, they are in our experience a localised, relatively small-scale event.

More usually, herbivores have less dramatic effects, causing clearings within forests, such as those formed by herbivorous fish such as Herring Cale (*Odax cyanomelas*) in New South Wales (see Chapter 24), or reductions in biomass — but not number — of kelp. An example of this latter effect is grazing by the arboreal sea urchin (*Holopneustes purpurascens*). This sea urchin lives in the canopy enmeshed within the fronds of kelp and other large macroalgae, and when in high densities it can cause the occasional death of kelp plants. However, densities are usually too low to cause mortality, and the effects of this herbivore are usually sublethal. Though mostly not studied, this is probably also true for the effects of the wide diversity of herbivorous snails, crustaceans and other invertebrates which live in and among kelp forests. These organisms may be very important, however, as grazers of microscopic kelp gametophytes, or of the very early stages of germlings or juveniles of other algae.

More generally, kelp and some other macroalgae are eaten by a wide variety of herbivores, which in turn are consumed by carnivores. Thus these plants directly or indirectly provide the basis for much of the overall productivity of reef communities. This includes many commercially important species such as rock lobsters, abalone and fish. Moreover, kelp play a more widespread role in these ecosystems through the so-called detrital food web. As described above, at various times of the year kelp and other macroalgae slough considerable amounts of tissue into the water column. They are also often torn from the bottom by storms. In both instances, the detached algal tissue becomes detritus — decaying organic material. This detritus is spread by currents and tides throughout the water column and up and down the coast, becoming food for both planktonic and benthic organisms that are not found in kelp forests. The resulting amount of decaying algal tissue present in nearshore systems can in fact be enormous, rivalling or exceeding the biomass or productivity of phytoplankton.

Most studies of kelp forests focus on the obvious — the kelp themselves or the associated larger fish or invertebrates. However, a final way in which kelp forests are analogous to terrestrial forests is that, like trees, individual kelp plants are a world unto themselves. Focusing exclusively on the obvious patterns and processes ignores a remarkable diversity of animals and plants living on or among the fronds of kelp and other macroalgae. These plants are often covered by other algae or sessile invertebrates which attach themselves to macroalgae (termed 'epibiota' or, more commonly, simply as fouling) rather than directly to the surface of the rocks. They are also home to small snails, crustaceans, worms and other mobile invertebrates, some of which feed on their hosts directly, and some on the small algae or invertebrates. These small herbivorous invertebrates which consume the epiphytic ('living on plants') algae are know as 'mesograzers', and can be thought of as the analogs of insects on terrestrial trees. Together, mesograzers and epiphytes occupy a world which pushes the limit of a diver's vision, but which is as diverse and complex as the overall forest itself. These epiphytic fouling communities experience many of the same ecological phenomena that the forest as a whole experiences but, because of the nature of the organisms involved, these processes occur on a much more rapid scale. The ecology of these fouling communities also has considerable applied implications. (See box on opposite page.)

CONCLUSIONS

Giant Kelps, Bull Kelp, the Common Kelp, Crayweed and other large brown algae like *Sargassum* and *Cystophora* create three-dimensional 'forests' that are occupied by other understorey macroalgae, sessile invertebrates, and mobile grazers and carnivores. Kelps and large brown algae thereby create both valuable habitat and food for many marine animals.

The processes that structure these habitats include disturbance by storms, grazing, and interactions between species for space and resources within the habitat. These processes interact so that within reefs there are cleared patches and gaps in the canopy within a forest. This mosaic of kelp and patches of other algae creates a diversity of habitat that in turn is occupied by many species of other algae, invertebrates and fish. Viewed superficially, this diversity makes for enjoyable diving and snorkelling, but kelp forests play a fundamental but poorly understood role in the ecology of temperate rocky reefs. They also have an emerging role in the development of industrial and pharmaceutical products. For all these reasons, the kelp forests of southern Australia are worth conserving and managing well.

Blacklip Abalone.
Venus Bay,
South Australia.
Rudie Kuiter

8 Blacklip Abalone

Paul McShane

Blacklip Abalone (*Haliotis rubra*) are some of the most abundant marine snails on rocky reefs in southern Australia. Their large muscular foot allows them to cling to rocky reefs in even the most exposed places. Blacklip Abalone are found around the southern half of the continent and Tasmania and occur in a range of rocky reef habitats and depths. They occur as far north as Coffs Harbour in New South Wales, around Tasmania, and west to Perth. Blacklip Abalone can be found in sheltered estuarine habitats and in some of the most exposed waters in the world, such as the west coast of Tasmania.

In South Australia, Blacklip Abalone are found in the undercuts of the limestone reefs and in the crevices and caves of granite reefs. In Victoria, large granite boulders form high-relief habitat with aggregations of Blacklip Abalone underneath the boulders rather than on the exposed upper surfaces. In New South Wales, Blacklip Abalone are found mostly in crevices and under boulders, although in the far south of the state they may be found on offshore reefs under dense canopies of Crayweed (*Phyllospora comosa*). These generalisations aside, Blacklip Abalone are found in a range of habitats. Unlike many other species of abalone, Blacklip Abalone occur over a wide depth range. Although they are most abundant in shallow water (< 10 m in depth), they can occur to depths greater than 60 m. Indeed, within their range, wherever seaweeds occur and rocky reef is available, Blacklip Abalone might be found.

Blacklip Abalone are most abundant in Victoria and Tasmania, and in these states, and in southern New South Wales, large aggregations are often found in the understorey community of Crayweed. This large brown alga grows to about 3 m in length. This association provides some interesting ecological insights into the interaction of abalone with algae and animals that share the subtidal reefs.

There are several colour variants to the normal Blacklip Abalone. Brownlip Abalone (*Haliotis rubra* var. *conicopora*) are found in Western and South Australia, and in Port Phillip Bay in Victoria in broadly the same areas as Greenlip Abalone. Genetically, they are not sufficiently different from Blacklip Abalone to be considered a true species but grow to a larger size (longer than 250 mm) and have raised pores on their shell. The shell is darker than the normal Blacklip Abalone shell. Brownlip Abalone support a small commercial fishery of about 20 t per year in Western Australia. Another colour variant of Blacklip Abalone is the so-called 'tiger blacklip'. These abalone have a distinctive light brown foot with dark striations but are otherwise indistinguishable from the more usual Blacklip Abalone. In New South Wales, 'tigers' are most common in the commercial catch south of Eden. In South Australia they more common in the gulfs and in Tasmania they are most common in the north of the state. Little else is known of the causes or distribution of this colour variant.

FOOD AND FEEDING

As do most abalone, Blacklip Abalone feed mainly on drift seaweed and, although they have clear preferences for different algae, their diets largely reflect the availability of seaweeds. In South Australia they feed mostly on red algae. In Victoria, Tasmania and New South Wales, Crayweed and other brown algae are more important as the red algae are less abundant. The species composition of drift seaweeds on a reef will be determined by algae present in the surrounding area and their vulnerability to being torn from the reef by storms. Both Crayweed and the Common Kelp (*Ecklonia radiata*) are abundant on reefs in southern Australia, along with Bull Kelp, and species in the genera *Sargassum* and *Cystophora*. All these algae are eaten by Blacklip Abalone. Interestingly, however, Common Kelp have high levels of tannin which appears to deter feeding in some instances.

GROWTH AND AGE

Growth is perhaps the most studied aspect of the biology of Blacklip Abalone; their shells can easily be tagged and they are sedentary so enough tagged abalone can be recovered to estimate growth rates. Blacklip Abalone grow quickly to near their maximum size before slowing, but the rate at which they grow and the maximum size attained varies among individuals on a reef and, at larger scales, among regions and states. Food availability, wave exposure and water temperature all contribute to variability in growth but the relative importance of these factors is not well understood.

Fast-growing Blacklip Abalone are easily recognised by the clean margin of the shell. Blacklip Abalone grow fastest in New South Wales, and the fastest reported growth is from Broughton Island, near Newcastle, where they may grow more than 40 mm in their first year of life. Growth rates in other parts of New South Wales and in other states are generally slower and may be less than half this rate. By the time abalone approach the minimum size at which they can be harvested, annual growth has slowed to less than 10 mm.

Differences are also found in the maximum size to which Blacklip Abalone grow — these sizes may or may not be related to the rate of growth. The largest Blacklip Abalone are from Tasmania where individuals up to 218 mm long and 3.6 kg in weight have been recorded. More usually, Blacklip Abalone grow to be between 115 mm and 140 mm before being harvested or dying of natural causes.

Differences in growth are reflected in differences in the size limits imposed on fisheries for Blacklip Abalone. For example, the minimum size limit on the west coast of Tasmania is 140 mm, 25 mm longer than that used in New South Wales. In other parts of Tasmania it is set at 132 mm. Smaller maximum sizes are also used to justify the 100 mm size limit used in Port Phillip Bay instead of that used in adjacent coastal waters (120 mm). The minimum size limit in South Australia is 130 mm. In some cases, perhaps because of differences in wave exposure or food availability, areas of coastal habitat in all states support abalone that do not grow to the legal harvestable size.

'Stunted' populations of abalone can be managed by applying a smaller than usual size limit for short periods in 'fish downs'. Such harvesting strategies have been successfully

▷ Pink encrusting coralline algae under Crayweed are a preferred settling surface for Blacklip Abalone. Point Home, Tasmania. *Matt Edmunds*

▽ Blacklip Abalone contrasting the distinctive 'tiger' colour variant (centre) with the more usual colouration. *Simon Talbot*

△ A tiny Blacklip Abalone grazing on coralline algae. This scanning electron micrograph shows both the smooth larval shell and the newly formed, ridged, adult shell. The abalone is 0.5 mm long and less than 2 weeks old. *Sabine Daume*

applied to Blacklip Abalone fisheries in South Australia, Victoria and in the Bass Strait islands in Tasmania. Benefits include access to abalone that would otherwise be unavailable to harvest, and improved growth and quality of abalone. In New South Wales, slow-growing Blacklip Abalone have wider shells than fast-growing individuals of the same length. Managers use this difference in a width-based size limit to allow commercial divers to harvest slower-growing abalone for short periods each year. Because the total harvest is limited each year, when these animals are harvested, fewer faster-growing Blacklip Abalone are harvested by commercial divers.

For most populations of Blacklip Abalone, marks visible in the shell cannot be used to age the abalone because marks are not laid down consistently. Some marks are growth checks, caused by a variety of disturbances, such as attack from predators, shell-boring parasites, freshwater inundation and so on. The episodic and unpredictable nature of these disturbances mean that counts of the marks are not reliable indicators of age. Indirect estimates of age can be derived from annual growth of tagged abalone. These studies indicate that Blacklip Abalone may live more than 30 years. Because of relatively high survival and slow growth, harvesting abalone at a comparatively large size can maximise yields from the fishery. There is a trade off between weight gain from allowing the abalone to grow to a large size and the loss from natural causes such as predation.

Minimum size limits are generally applied to balance harvesting with the capacity of the population to replenish itself. If the size limit is set too high, then only a small proportion of the available stock will be available for harvesting and the generation of economic wealth will be adversely affected. This is recognised explicitly in the fish downs of stunted abalone stocks. If the minimum size limit is set too low then immature abalone could be exposed to fishing and the risk of overfishing will be unacceptably high. Hence the many different size limits for Blacklip Abalone (three in one zone of Victoria alone), ranging from 100 mm (shell length) in Port Phillip Bay to 140 mm in Tasmania.

REPRODUCTION

Blacklip Abalone become sexually mature at about four years of age. However, because growth rates vary, size at maturity in Blacklip Abalone varies. In New South Wales, they mature when they are about 90 mm long. Males and females exist in approximately equal proportions in Blacklip Abalone. Females produce up to 6 million eggs/year. Because abalone reproduce by casting eggs and sperm directly into the water, the aggregating tendency of Blacklip Abalone enhances successful fertilisation of eggs. Blacklip Abalone can form dense aggregations of more than 100 individuals, but aggregation size varies with habitat.

Larvae of Blacklip Abalone remain in the plankton for less than five days and can be retained in the area in which they were spawned by topographic features and the canopy of forests of Crayweed and other large brown algae. This retention reduces the dispersion of larvae to distant places and less certain futures. The understorey community associated with Crayweed provides suitable surfaces for larvae to settle and juveniles to grow on. The larvae of Blacklip Abalone settle preferentially, but not exclusively, on crustose coralline algae. Factors that determine the distribution and composition of coralline algae therefore have the potential to influence the growth and survival of abalone. Such factors include the presence or absence of grazers such as other marine snails and sea urchins. Grazers are important in maintaining coralline surfaces free of overgrowth by other seaweeds.

The great abundance of Blacklip Abalone on some reefs suggests that they can be important ecologically. The ecology of the abalone has obvious implications for commercial abalone fisheries but is surprisingly poorly understood. The juvenile habitat, typically in shallow water, is the source of fisheries production. High rates of settlement occur on coralline surfaces in dense forests of Crayweed (more than 10 000 abalone/m^2 have been recorded). Such high rates of settlement probably are caused by the larval retention, attenuation of water flow and the availability of habitat suitable for the growth and survival of post-larval abalone. Because of the wide distribution of Crayweed, and its dominance on subtidal reefs to 10 m depth, habitat types used by adults and juveniles often overlap, providing a clear means for populations to recover from fishing. In other abalone species, in which the adult and juvenile habitat can be greatly separated, recovery from fishing can be very slow. In Victoria and New South Wales, large differences in juvenile settlement have been recorded among years. These differences are episodic events that reflect the prevailing oceanographic conditions at the time of reproduction.

ECOLOGICAL INTERACTIONS

Blacklip Abalone, particularly those in their first year of life, fall prey to a wide range of predators on rocky reefs. Notable among these are Port Jackson Sharks (*Heterodontus portusjacksoni*), wrasses such as Eastern Blue Groper (*Achoerodus viridus*), Crimson-banded Wrasse (*Notolabrus gymnogenus*) and the Blue-throated Wrasse (*N. tetricus*), and octopuses, stingrays and rock lobsters. Attacks from most predators such as octopuses and starfish elicit a defensive response in the abalone — clamping firmly to the reef in the case of octopuses or actively escaping from seastars. Blacklip Abalone gain some refuge in size as they grow.

Humans are now one of the most important predators of Blacklip Abalone. In New South Wales, mathematical modelling suggests that the present population is about 40 per cent of what it was before commercial fishing. In other states this percentage is unknown. Blacklip Abalone have all but disappeared from many reefs, particularly around major population centres. The impact of these reductions on the ecology of shallow rocky reefs in southern Australia is poorly understood.

Abalone offer good opportunities for study in their natural habitat. They live in shallow, accessible coastal habitats and attributes of their ecology have been studied in most states. In New South Wales, Blacklip Abalone compete with Black Sea Urchins (*Centrostephanus rodgersii*). Grazing by this sea urchin maintains about 50 per cent of the area of nearshore reefs in southern New South Wales free of large seaweeds (see Chapters 2 and 15). When sea urchins are removed, densities of small Blacklip Abalone increase by an order of magnitude. These observations confirm the long-held view by commercial abalone divers that sea urchins displace abalone on reefs. However, the nature of the interaction is likely to be more complex. The ecological consequences of removing all large brown algae such as Crayweed may have a greater impact than simply via direct competition for food.

FISHERIES

Blacklip Abalone support important commercial fisheries in all the southern states. Slightly less than 5000 t of Blacklip Abalone are caught each year in Australia, of which, 2510 t is harvested from Tasmania. The total Australian harvest of Blacklip Abalone, represents about 40 per cent of the total world catch of abalone, making Blacklip Abalone the most important of the world's commercial species. Because of the high price paid on Asian markets, abalone is one of Australia's most valuable seafoods.

All commercial Blacklip Abalone fisheries are managed by a mixture of minimum size limits, total allowable catches, and seasonal and area closures. The persistence of these commercial fisheries since the 1960s attests to the apparent resilience of this species to prolonged and intense fishing. The collapse of abalone fisheries in North America and the decline of many others suggests, however, that prudent and flexible management is required to ensure the sustainability of these important fisheries.

Greenlip Abalone on a typical limestone reef. Robbins Island, Tasmania. *Simon Talbot*

9

Greenlip Abalone

Scoresby Shepherd

Greenlip Abalone (*Haliotis laevigata*) are distributed across southern Australia from Cape Leeuwin in Western Australia to Flinders Island in Bass Strait. The ocean currents, tides and upwellings along parts of this southern edge of the continent bring cool, rich subantarctic waters close inshore and promote a profusion of algae on the rocky reefs and islands that sustain Greenlip Abalone. Greenlip Abalone share these reefs with a diversity of other animals including predators, competitors and parasites and have evolved an impressive suite of life history features to cope. As a measure of their success, Greenlip Abalone are the dominant herbivore on reefs in central South Australia. In this chapter I describe the natural history of Greenlip Abalone over a range of scales from individuals to groups of populations. Finally, I describe in greater detail food and feeding, defences against predators and reproduction, growth and mortality.

Greenlip Abalone are typically found on low-relief limestone reefs and to depths of at least 30 m. To a lesser extent they occur at the bases of cliffs and at the margins of reefs in the lee of islands or on other shores protected from the oceanic swells rising from the Southern Ocean. Within this broad range, however, Greenlip Abalone are not found in all places with seemingly appropriate habitat. Their populations are isolated from each other and often separated by 20–50 km of sand or even rocky reef. The largest populations or groups of populations are found off Eyre Peninsula and in Spencer Gulf, South Australia. The largest population is at Tiparra Reef, Spencer Gulf, which extends over 30 km^2 of reef, and produces about 100 t of Greenlip Abalone each year for the commercial fishery.

Groups of populations are important for their persistence on many separate reefs. Clusters of populations are 'held together' by the extent of suitable rocky reef and the distance

larvae are transported among them. If enough larvae reach a given site, a local population will establish itself and will in turn contribute larvae to perpetuate itself and nearby populations. Around islands, populations are typically small and largely restricted to the lee where the eddies in currents trap larvae close to shore. Similarly, many bays contain isolated populations because their larvae are retained by tidal currents. However, the greatest concentrations of Greenlip Abalone, such as those off Cape Leeuwin, Western Australia, in Thorny Passage and Tiparra Reef in Spencer Gulf, and those in Franklin Sound in the Furneaux Group, Bass Strait, occur on open and extensive areas of reef. Here, the twice daily reversing tidal currents over the 4–5 days of larval life still retain many of the larvae.

Greenlip Abalone are not simply scattered about reefs by chance and circumstance but are organised into loose aggregations at a scale of tens to hundreds of metres. The aggregations tend to occur in crevices and where there is an abundant food supply. Movement is risky for abalone so they move at night when predators are not active and only when there is a prolonged shortage of food. In some places where predation is apparently low, Greenlip Abalone abandon the safety of crevices to gain a better site for feeding in large aggregations.

△ Greenlip Abalone on a reef. Victor Harbour South Australia. *Rudie Kuiter*

During the spawning season, if not already in clumps, they aggregate with nearby individuals and prefer places of strong current where their eggs and sperm have a high chance of dispersing.

The habitat of Greenlip Abalone changes with size. For the first six months of their lives Greenlip Abalone live on encrusting red algae, the crustose corallines. Even in this microhabitat the tiny abalone has subtle preferences, only recently discovered. There are at least four growth forms of crustose corallines. One is warty, another is flaky in appearance, a third has smooth rounded protuberances, the fourth is smooth. Each form has its own suite of diatoms living on it. Newly settled abalone have strong preferences for particular diatoms and growth forms of corallines and these preferences change as abalone grow. After six months, small abalone move to a crevice habitat for several years and then, as sexual maturity approaches at 3–4 years of age, they assume a more mobile and exposed existence. This is often spatially separated from that of juveniles, so the gradual movement of young abalone within the reef system is common. Migration of several hundred metres in a year has been recorded.

FOOD AND FEEDING

Greenlip Abalone eat different foods as they grow. Immediately after settlement onto the reef Greenlip Abalone feed on diatoms and bacteria on crustose coralline algae. For six months tiny Greenlip Abalone rarely move off the crustose coralline except to explore nearby surfaces in search of food. Their diet at this size is increasingly the surface tissue of crustose corallines which they rasp from the rock surface. At about 1 cm length they leave the coralline algae and their food changes first to the low algal turf of filamentous algae. With increasing size they browse on larger drift seaweed which is trapped among rocks. The happy circumstance for small Greenlip Abalone is that the same currents and eddies that transport larvae into suitable habitat, also brings drift algae. Hence in the early years of life small Greenlip Abalone are assured of plentiful food but as they grow their food supply may become limiting. In these situations Greenlip Abalone may move in the direction of approaching swell or stronger current, where they are likely to find more food.

Greenlip Abalone, like other abalone, can greatly extend the forepart of their foot. As swell increases the animal rears up on the back of the foot, extends the fore-foot and waits for a piece of drifting weed to arrive. This it grasps with extraordinary agility as seaweed is swept back and forth by the swell. Large Greenlip Abalone do not feed indiscriminately, but select particular kinds of algae. Brown algae, often rich in defensive tannins, are distasteful and tough, and Greenlip Abalone avoid them. Delicate red algae, common on reefs within the range of Greenlip Abalone, are preferred. The periodic strong swells common on exposed reefs tear off algae and produce algal drift, and in these conditions, and in the ensuing days, when the swell moderates and transports the drift algae along the bottom, the Greenlip Abalone traps abundant food.

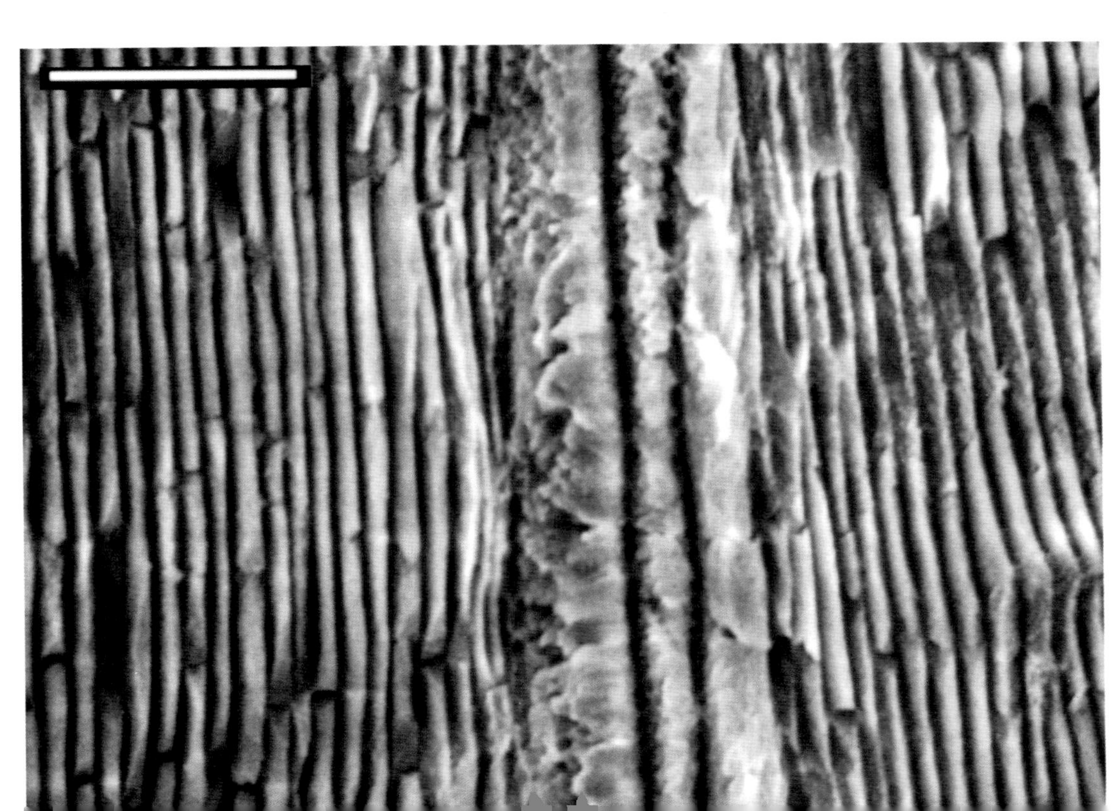

▷ Scanning electron micrograph of the shell of a Greenlip Abalone showing the fine brick-like nacre on either side of a ring of biopolymer. Scale bar is 5 µm *Scoresby Shepherd*

DEFENCES

In a world full of predators, defences are critical to survival, and abalone have two kinds of defence: behavioural and structural. Even sedentary animals such as abalone have a repertoire of behaviours to help them survive. The most notable is the well-developed flight response to starfish and some predatory molluscs. This can be demonstrated by placing the large seastar *Coscinasterias muricata* near a small abalone. The abalone first protrudes many sensory tentacles, which wave about, presumably sensing chemically the proximity of the seastar. Within a second or two, the abalone twists its shell violently back and forth, and then moves rapidly away. Greenlip Abalone larger than about 8 cm do not attempt to flee but clamp down on the reef and exude a white substance from special glands that sometimes, but not always, appears to deter interest by the seastar. Why do large abalone adopt a different defence from small ones? One possible reason is that flight itself induces other hazards. Wrasses are often seen following predators such as stingrays and seastars, and readily steal or scavenge pieces of abalone left by them. Wrasses will attack an exposed and fleeing abalone if given a chance. A better strategy for large Greenlip Abalone is therefore to clamp down and try to outlast the attack. This strategy, although often successful against seastars, can be fatal against octopus attack. The octopus simply spreads its mantle over the abalone and slowly suffocates it. In a short while the abalone's foot muscle relaxes and the octopus can pull it off the rock and transport it to its lair.

For small Greenlip Abalone the best defence is to live in a crevice or under a boulder where its main predators, crabs and wrasses (see Chapter 21), cannot attack easily. However, the small abalone have to move about within their micro-habitat to search for food. Crabs use their claws to prise a small abalone off the rock, and in doing so they nearly always chip the edge of the shell. Numbers of empty chipped shells have been used to estimate the numbers of small abalone taken by crabs. Wrasses, mainly the Blue-throated Wrasse (*Notolabrus tetricus*) and the Senator Wrasse (*Pictilabrus laticlavius*), ram the shell, reducing its suction on the rock and in the same action pulling the abalone off the rock. At one site intensively studied for 15 years, the annual mortality rate of 1–2 year old Greenlip Abalone was mostly in the range of 40–70 per cent; of these deaths crabs accounted for about a fifth and wrasses for the rest. At this same site a group of seals established a haul-out colony during the study, and substantially reduced the wrasse population; the mortality rate of small abalone subsequently declined to half its previous level.

The most important defence for the abalone is its shell. This complex laminated structure simultaneously provides strength, hardness and toughness. The shell is made of layers of nacre alternating with rings made of prismatic elements running cross-wise and embedded in biopolymer. The composite nature of abalone shells make them twice as hard and many times as tough as their constituents. Such a hard shell is needed to defend abalone against larger predators such as stingrays and blue groper, both of which try to crush the shell during capture. Stingrays typically set their lower jaws against the base of a shell near the spire for leverage and their upper jaw on the flatter upper surface. Then, vigorously flapping its wings, the ray tries simultaneously to crush the shell and prise it from the rock. Blue groper (genus *Achoerodus*) simply ram the shell by brute force with their teeth and attempt to fracture it. Many Greenlip Abalone can be seen alive with the tell-tale scratch marks of stingray teeth a proof of the effectiveness of the shell as a defence — but the number of broken shells on the reef is proof of final failure.

Although less common, drilling whelks also attack Greenlip Abalone but often not fatally. These small whelks take 5–7 days to bore through the abalone shell, extract juices for a day or so and then withdraw when the abalone exudes a deterrent brown material around the drill hole and begins to repair the shell. Another less obvious form of attack comes from boring sponges and worms that, with the passage of years, gradually weaken the shell so that eventually, if not taken by a diver, the abalone will fall easy prey to a stingray.

Not all co-habitants of Greenlip Abalone are harmful. In some places a clingfish (Family Gobiesocidae) lives under the shell, deriving protection but causing no harm. In other places a small amphipod is a tenant, and may be seen diving for cover under the shell as a diver or ray approaches, whereupon the Greenlip Abalone swiftly clamps to the rock below. It is conceivable but not proven that the amphipod gives a warning to the abalone of a predator's approach.

▷ Juvenile Greenlip Abalone on a reef. Kangaroo Island, South Australia. *Ken Hoppen*

GROWTH AND AGE

The food supply of an abalone is often limited and may change seasonally depending on the habitat. The energy provided by this available food must be divided between the abalone's needs for growth, reproduction and maintenance. Abalone have evolved to divide their food resources best for survival. For example, abalone in a place subject to high predation from stingrays (perhaps due to a scarcity of crevice space) may need to invest much more energy into growth early in life so that they can grow rapidly to a large size and so become less vulnerable to predation by rays. On the other hand, in other conditions where shells become parasitised and weakened after a few years, then a better strategy is to grow slowly and invest much more energy into producing eggs early in life to keep the population going. Despite these types of responses to local conditions, there are some common patterns in growth of Greenlip Abalone. In many populations growth rates of 20 mm a year for the first 5–6 years are common. After that, growth rate slows, and depending on the food supply, Greenlip Abalone reach a maximum size of 150–200 mm after 10–15 years.

In a few places limpets live on the shells of Greenlip Abalone and scrape the algae that grow there. While not directly damaging the abalone, the constant abrasion of the shell by the limpet gradually erodes the shell's outer layer and exposes the underlying shiny nacre, hence the popular name 'shinybacks' given to them by divers. The unfortunate shinyback must now spend more energy laying down fresh nacre on the inside of its shell to replace the fast-eroding outside. With little energy left to devote to growth, the shinyback typically does not grow to be as large as abalone without limpets.

The longevity of abalone depends, of course, on their mortality rate throughout life. In populations that are unexploited, where approximately 15–20 per cent die every year after about three years of age, the oldest individual likely to be found would be about 25 years old. Natural mortality is lowest, possibly only 10 per cent a year, in places of very high swell at the base of cliffs, and here 30-year-olds have been sometimes found. On more open reef, in calmer conditions, as many as 40 per cent may die each year, largely to stingrays. For most Greenlip populations, fishing is the major source of mortality for large animals.

REPRODUCTION AND EARLY LIFE HISTORY

A large female Greenlip Abalone can produce up to 10 million eggs a year at some sites, although at other sites it may be as little as one million. Greenlip Abalone become sexually mature at the beginning of the fourth year of life but substantial numbers of eggs are not produced for another two years. Spawning occurs annually from October to December and is synchronous in a population. A cascade of external events cue the developmental changes leading up to and triggering spawning. Increasing day length after mid-winter causes uniform development of the gonads and the slowly increasing sea temperatures during spring will bring them to maturity at about the same time. Finally, it is believed that strong tides or the moonlight around the full moon causes epidemic spawning.

Abalone, like other primitive gastropods, shed sperm and eggs into the water where fertilisation takes place. For successful fertilisation, males and females need to be close to each other and to release gametes at the same time (apparently achieved by males spawning first followed by the females). When large clusters of abalone spawn, nearly 100 per cent of the eggs are fertilised, compared with only a small percentage when few animals are clumped. The advantages of a whole population spawning at the same time are less obvious, but may relate to the lower risk of mortality when a large number of larvae settle together in the crustose coralline habitat. The loss of reproductive success when animals are distant from each other is called the 'Allee' effect after the ecologist who first described it. A decline in density reduces by that amount the number of eggs produced in the population, and this reduction is exacerbated by the reduced fertilisation of eggs produced by increasingly isolated females.

Some of the fertilised eggs sink to the seabed and others slowly drift away into deeper water. Some eggs fall into crevices between boulders where they are eaten by sea anemones, bryozoans and ascidians. Before dawn the surviving fertilised eggs have hatched into trochophore larvae and are at the mercy of the swell and currents. Some are eaten by planktivorous fish. After a day the trochophore larvae develop into veliger larvae and drift for four more days with tide and current. On the fifth day they begin to sink and periodically touch the reef, testing the surface. If they happen to land on crustose coralline algae they stay there and soon transform into a form that resembles the adults. Of the many millions of eggs fertilised, only a very small fraction survive to settle on the reef; the others are lost at sea. Of those that sought to settle some are captured by worms which sweep the coralline surface with their tentacles. Others fall prey to sessile predators such as sea anemones or ascidians.

EXPLOITATION

Commercial fisheries for Greenlip Abalone exist between Cape Leeuwin in Western Australia and Flinders Island, Tasmania. The total commercial harvest is currently a little over 1000 t but has declined by 25–50 per cent since commercial exploitation began in 1965. The main reason for these declines is that small populations are more vulnerable to overfishing than larger ones for the simple reason that a small population, without a safety buffer in numbers, always has a greater risk of extinction that a large one. During the three decades of fishing the smallest populations have largely disappeared and many other slightly larger ones have declined to varying extents. The challenge facing abalone biologists and fishery managers is to halt further declines of existing stocks, and to take restorative measures for those that have declined. Sustainable harvesting of Greenlip Abalone for profit and recreation requires both vigilance to detect early signs of decline and positive steps to restore what has been lost.

Giant Cuttlefish.
Bermagui,
New South Wales.
Mary Malloy

10 Octopuses & their relatives

Mark Norman

THE HIDDEN INHABITANTS OF ROCKY REEFS

A wide range of octopuses, squids, cuttlefishes and their relatives (cephalopods) live on rocky reefs in temperate Australia. These animals are, in many ways, the invisible inhabitants of these reefs because of their cryptic behaviour. This behaviour may account for our fragmentary knowledge of the diversity and natural history of most Australian cephalopods. Their soft bodies and lack of armour make them highly vulnerable to attack by their many predators (predominantly fish) and, as a consequence, the group has evolved diverse and efficient means of avoiding detection and escaping attackers. These behaviours range from excellent camouflage abilities, to nocturnal activity patterns in many species and/or hidden existences deep in the cover of ledges, crevices, rubble or kelp plants.

A dive on a southern Australian rocky reef during the day is unlikely to find many live cephalopods. At most, you may encounter a cuttlefish camouflaged amongst algae or see the eye and suckers of a large octopus deep in a crevice. A scatter of clean crab carapaces or intact bivalve shells indicate a nearby octopus lair, potentially hidden behind a wall of rocks. The shell remains are the discarded kitchen scraps of a night-active octopus. A shiny bivalve shell under a ledge identifies a temporary octopus feeding station, where the foraging animal has paused to consume one of the night's catch. Groups of white eggs the size of golf balls on the underneath of a ledge indicate the site of a cuttlefish spawning. Mops of fleshy white egg strings swaying amongst the algae mark a squid spawning site. A lobster pot with an empty lobster shell indicates an octopus ambush, sure to frustrate the lobster fisher.

A night dive on a southern rocky reef is more likely to encounter active cephalopods. Large foraging octopuses move along ledges

or over broken ground. Small blue-ringed octopuses may be found foraging amongst weed and rubble. Other cephalopods move in over reefs at night from adjacent habitats. Southern Calamari Squid (*Sepioteuthis australis*) may approach to feast on the small fishes and shrimps attracted to dive lights. Other small rounded squids, normally found associated with seagrass beds or sand substrates, also move over rocky reefs at night to forage. Whether day or night, some cephalopods are unlikely to ever be seen by divers. These are the tiny pygmy octopuses which spend their entire lives under cover, in the safety of kelp holdfasts or deep under ledges. Such species can be only one gram with an armspan of less than 3 cm at maturity.

DIVERSITY

In contrast with many of the fishes, few reef cephalopod species have distributions that span southern Australia from northern New South Wales to Western Australia. Those that do include the Giant Cuttlefish (*Sepia apama*) and possibly the Southern Calamari Squid. There is some question over the latter species because recent genetic studies suggest there may be regional differences in this squid from east to west, potentially warranting distinct species status. Although the great majority of reef cephalopods appear to be found only in certain regions, care should be taken in making generalisations because the cephalopods of whole regions such as Western Australia are still poorly described.

Similar-sized species appear to fill similar ecological roles or niches across the southern coast. All rocky reefs possess at least one large octopus species. In New South Wales and southern Queensland, this is the common Sydney Octopus (*Octopus tetricus*) which is often seen in the mouth of lairs showing its distinctive orange-edged arms and eyes with white irises. In comparable warm-temperate latitudes of Western Australia, this octopus is replaced by a related but undescribed species (*Octopus* sp.) that is currently treated in the literature as *O. tetricus*. It is likely that these eastern and western species share common ancestry but were isolated by cooler southern waters during past glacial periods. Similar sister species between east and west also occur in many other groups of marine invertebrates and fishes. In the cooler waters of South Australia, Victoria and Tasmania, the 'resident large octopus' is the Maori Octopus (*Octopus*

△
The Frilled Pygmy Octopus is one of the small, cryptic octopuses found on reefs. Portsea, Victoria.
Mary Malloy

◁
The Maori Octopus is the largest octopus speciesfound in southern Australia. Portsea, Victoria.
Mark Norman

▷
A Giant Cuttlefish above the reef. Jervis Bay, New South Wales.
Kevin Deacon

△
Velvet Octopus lack an ink sac and probably originate from a deep-water ancestor. The one shown here is tending her eggs. Esperance, Western Australia.
Mark Norman

A Club Pygmy Octopus showing its club-like modified arm tip on the third right arm. Portsea, Victoria.
David Paul

▷
A female Southern Blue-ringed Octopus carrying her eggs. Port Phillip Bay, Victoria.
David Paul

◁ The Sydney Octopus is the most common octopus in southern Australia. Sydney, New South Wales.
Mark Norman

maorum). These octopuses are the largest in Australia and have been recorded with arm spans longer than 3 m and weighing more than 10 kg.

In South Australia, an unusual octopus is found on shallow rocky reefs around Spencer Gulf and Gulf St Vincent. The Velvet Octopus (*Grimpella thaumastocheir*), appears to show taxonomic affinities with deep-water species. As with many deep-sea octopuses, this shallow-water species lacks an ink sac. Little is known of the biology of this species and even less is known of temperate Australian pygmy octopuses. At this stage, only the species of Victoria and Tasmania are well known. These include the Club Pygmy Octopus (*Octopus warringa*) and the Frilled Pygmy Octopus (*Octopus superciliosus*).

In all temperate waters around Australia, medium-sized octopuses (weighing less than 1 kg and having arm spans less than 50 cm) may be seen on rocky reefs, having moved in from by adjacent seagrass or soft-sediment habitats. In New South Wales these include the Hammer Octopus (*Octopus australis*). In South Australia, Victoria and Tasmania, these include the Southern Keeled Octopus (*Octopus berrima*), the Pale Octopus (*Octopus pallidus*), the Southern White-Spot Octopus (*Octopus bunurong*) and the Southern Sand Octopus (*Octopus kaurna*). There is still little information available for Western Australian waters.

Blue-ringed octopuses also occur on all rocky reefs of southern Australia. These smaller animals only reach arm spans of around 20 cm and weights of less than 100 g. These distinctive octopuses are famed for their venomous nature and are recognised by iridescent blue markings over the upper surfaces of the arms, head and body. At least two blue-ringed octopus species occur in temperate Australian waters. In New South Wales and southern Queensland, the Blue-lined Octopus (*Hapalochlaena fasciata*) occurs on intertidal and subtidal rocky reefs. This species is the only Australian blue-ringed octopus with blue lines on the body, lacking the rings of various sizes found in other species. In Victoria and Tasmania, this species is replaced the Southern Blue-ringed Octopus (*Hapalochlaena maculosa*) which has tiny (1–2 mm) blue rings on the body.

The cuttlefishes of temperate rocky reefs are harder to identify. The primary diagnostic tool is the cuttlebone and several species are known only from cuttlebones found on

△
The Southern Dumpling Squid sometimes forages over rocky reefs, moving in from adjacent sand and seagrass habitats. Edithburgh, South Australia.
Mark Norman

◁
Two male Giant Cuttlefish in display combat. The reason for their combat, a female laying eggs, is obscured from view. Spencer Gulf, South Australia.
Mark Norman

▷
A Southern Calamari Squid devouring a fish attracted by the photographer's lights.
Mark Malloy

◁ A pair of Southern Calamari Squid guarding their eggs. Jervis Bay, New South Wales. *Paul Baumann*

beaches. The largest cuttlefish in the world occurs in this region and is the easiest to identify in the field. The Giant Cuttlefish (*Sepia apama*) reaches sizes of 1 m total length with a cuttlebone up to 50 cm long. It occurs from southern Queensland to Coral Bay, Western Australia. It can be recognised by its large size, or, for smaller animals, by two rows of three flattened lobes over each eye, which can be raised as tall branched flaps. New South Wales divers will be familiar with the Small Reaper Cuttlefish (*Sepia mestus*), a robust red species typically less than 15 cm long which frequently shows a pair of black spots towards the rear of the body. It is often seen at the back of crevices at places such as Jervis Bay and Montague Island. Many other cuttlefish species occur in deeper habitats, potentially on rocky reefs, while others move in over reefs from adjacent seagrass and soft-sediment habitats. Diagnostic characters enabling identification in the field are still being developed.

The following small rounded squids, typically found on seagrass beds or soft sediments, can also be encountered on rocky reefs at night: the Southern Dumpling Squid (*Euprymna tasmanica*), Southern Bobtail Squid (*Sepiadarium austrinum*), Striped Pyjama Squid (*Sepioloidea lineolata*) and Southern Pygmy Squid (*Idiosepius notoides*).

CEPHALOPODS AS PREDATORS AND PREY

All cephalopods are voracious fast-growing predators, preying on a wide range of live fish, crustaceans, molluscs and other reef animals. The different major groups employ different hunting and prey manipulation methods. The octopuses of temperate reefs typically forage at night, using their arms and sensitive suckers to feel for prey. All octopuses have large salivary glands that produce potent saliva, used to quickly immobilise their prey. Some species target active prey such as fish and crustaceans. These octopuses forage by probing arms down crevices and burrows or by enveloping smaller rocks in their flared webs, using the free arm tips to flush prey into the webs and waiting suckers. Other octopuses collect shelled prey such as bivalve and gastropod molluscs that are carried back to the lair for drilling and eating. Such octopus species use their toothed tongue (radula) to drill through shells and poison/relax the occupant in order to gain access.

The salivary glands of squids and cuttlefishes are less well developed and do not appear to have the immobilising toxins of the octopuses. These cephalopods are visual predators. They tend to approach and attack prey from behind in order to quickly bite through a dorsal nerve cord, thus paralysing or killing their prey. The excellent camouflage skills of many cuttlefishes allow them to ambush prey amongst algae, while the faster squids will attack free-swimming fishes and crustaceans.

Cephalopods are present in the diet of many temperate reef predators, particularly fish. The teeth marks frequently found in cuttlebones washed up on the shore indicate the prevalence of cuttlefish in the diets of larger fishes and dolphins.

REPRODUCTION

Temperate reefs act as breeding grounds for many cephalopods. Female octopuses excavate lairs or wall themselves into crevices with rocks held on their suckers, laying their eggs singly or in strings on to the roof of these refuges. Blue-ringed octopuses carry their eggs in strings in the webs of their rear arms, using the cover of rocky reefs for refuge. Some soft-sediment octopuses and squids move to reefs in order to deposit and/or brood their eggs in more protected places, such as large embayments and deeper reefs. Mobile squid such as the Southern Calamari Squid anchor strings of eggs to the bases of seaweeds or under rocky ledges, leaving the eggs to fend for themselves. Evidence of at least small breeding aggregations of this squid are found as multiple egg masses at a single site.

The sexes are separate in cephalopods. Mating consists of males transferring encapsulated sperm packages (spermatophores) to the female, who stores it in various locations (depending on the species), ranging from the oviducts or ovary, to special sperm receptacles in the mantle cavity or around the mouth. The sperm can even be injected under the skin in some species. Immature females can mate and accept sperm packages which can be stored for long periods of time, up to 10 months in some species until they become sexually mature. Except for some deep-sea families, all familiar cephalopods are short-lived (most less than two years) and fast growing, with a single spawning event after which the female dies. This spawning event is typically of short duration, however several groups have protracted or pulsed spawning.

Squid and cuttlefish lay eggs singly or in strings that are encapsulated in a gelatinous coating, the eggs typically being laid in crevices or under rocks. The eggs of these cephalopods are not tended, parental brooding only being known in octopuses. The eggs of octopuses are tended, cleaned and oxygenated with jets of water by the female until hatching.

The most spectacular breeding aggregations found on the rocky reefs of southern Australia are those of the Giant Cuttlefish. At sites where there is limited rocky reef amongst large areas of seagrass and mud substrates (such as inside Spencer Gulf in South Australia), mature cuttlefishes gather in their thousands in autumn for mating and egg-laying. At these events, large males display to females and repel rival males using dramatic light shows of moving zebra stripes coursing over their bodies while webs on the arm tips are extended as long flowing white banners. Mating occurs face-to-face and the male deposits sperm packages in a special pouch under the female's mouth, sometimes using strong jets of water to flush the sperm of previous males from this receptacle before mating.

The female giant cuttlefish then lays single large eggs deep under boulders and in crevices, passing each egg over the sperm stores to fertilise them. The male 'mate guards' the female, standing vigil to ensure that no other males sneak in and replace his sperm. Males show ritualised aggressive displays where they pair off for intensive colour shows that determine which male is the largest or most impressive. These stand-offs rarely deteriorate to physical combat but some males show some bites or skin tears caused by the suckers. The breeding season ends with the death of females and larger males. At this time of year giant cuttlefish in poor health are a familiar sight underwater. The mass death of cuttlefish at the end of the breeding season can lead to the surface waters and nearby beaches being strewn with cuttlebones.

Dolphins and large predatory fish appear to take a heavy toll at these breeding aggregations as shown by the number of bones washing up with rows of deep punctures caused by large teeth. Dolphins have been seen to streak through aggregations of cuttlefish, leaving only explosions of cuttlefish ink in their wake. It is most likely that similar aggregations of other species of cuttlefish occur on other deep rocky reefs, still to be witnessed by divers or remote cameras.

INTERACTIONS WITH HUMANS — FISHERIES AND POISONOUS OCTOPUSES

Many species of temperate Australian cephalopods are of current or potential economic value. Octopuses and squids are exploited in both target and bycatch fisheries off New South Wales and Tasmania. Surveys of octopus stocks in Western Australia in the late 1970s estimated annual catches in excess of 30 000 tonnes in pot fisheries. Southern Calamari Squid are harvested in both recreational fisheries and in the bycatch of trawl fisheries. Giant Cuttlefish in South Australia have been harvested as bait for tuna line fisheries for many years and are now receiving increasing attention for human consumption. As fisheries based on fish attain or exceed maximum exploitation, and the seafood preferences of Australians become more diverse, the profile and value of cephalopods in Australia is likely to increase. Elsewhere in the world these animals are highly prized, attaining market prices of over US$50/kg for certain species. The world catch of cephalopods now exceeds 2 million t annually. There are some suggestions that the rapid growth of these animals mean that they are like 'weed' species, potentially taking over the ecological roles once occupied by longer-lived fish stocks.

Cephalopod fisheries, like some other invertebrate fisheries, have to be managed carefully. Recent expansion of the fishery targeting Giant Cuttlefish spawning aggregations may have a significant negative effect on the recruitment of the cuttlefish in regions such as Spencer Gulf in South Australia. The annual catch of this species from a single reef has risen from 2 t to more than 200 t in less than three years. This cephalopod species lays few (~100) large eggs (~40 mm long) and is less likely to sustain heavy fishing pressure than more fecund species in which females produce hundreds of thousands of tiny eggs (1–2 mm long).

Cephalopods also have negative economic impacts. Predation of rock lobsters by octopuses takes a heavy toll on these extremely valuable Australian fisheries. Lobster fishers report that up to one in three lobsters are lost to octopuses; these estimates are derived by the number of lobster carapaces left behind in lobster pots. The introduction of escape hatches for juvenile lobsters has significantly reduced capacity to catch the culprit octopuses, which use these trapdoors to escape. Where pots without trapdoors are used, octopuses are raised with the pots, killed and kept for bait or discarded. A recent Western Australian study estimated around 300 t of octopuses were discarded from the lobster fishery each year from one region alone; the main culprit species in Victoria, Tasmania and South Australia is the Maori Octopus.

One group of cephalopods on temperate Australian reefs poses a direct threat to humans. The venomous blue-ringed octopuses have been responsible for a number of human deaths and near-death incidents. The venomous saliva of these small octopuses contains tetrodotoxin, which is produced by bacteria harboured in the salivary glands. This powerful toxin acts by paralysing muscles over which we have conscious control ('voluntary muscles'), such as those used in breathing and limb movement. 'Involuntary muscles' such as the heart, iris and gut do not appear to be affected, so the paralysed victim may be fully conscious but incapable of breathing or moving. Mouth-to-mouth resuscitation may keep the patient alive until the toxin wears off. Recovered bite victims speak of the trauma of people discussing their demise while standing around them. Despite their deadly potential these small animals are relatively docile and are much more interested in escape than attack. They are common in southern Australia and are well camouflaged when foraging. It is only when they are harassed or feeding that the brilliant iridescent blue rings are flashed as a warning. In the few recorded deaths, the animal was being held in the hand or on the skin and being prodded to display its blue warning colours. No pain was felt and a pair of tiny bruises was the only evidence of a bite.

CONCLUDING REMARKS

There is still relatively little known of the Australian cephalopod fauna. A large proportion of Australian species have been recognised as new to science only in the last ten years and the count continues to rise as more work is done. Australian waters are proving to contain the highest diversity of cephalopod species reported for any region of the world. With the relatively recent arrival of SCUBA diving techniques and equipment, and the use of underwater still and video photography, information on the natural history, biology and impacts of these animals is slowly coming to light. There is still much to be learned of these fascinating, alien creatures. Detailed information on composition, distributions, stocks and biology will aid effective management and protection of these animals, both for exploited species and those potentially impacted by other human activities.

Jelly Blubber
over reef in
Jervis Bay,
New South Wales.
Joachim Ngiam

11 Jellyfish

Kylie Pitt and Michael Kingsford

Jellyfish are a fragile and beautiful sight when viewed underwater, yet their reputation for sometimes causing painful, and occasionally serious, injuries has made them an animal that is feared and avoided. Jellyfish are common in the temperate waters of southern Australia. Some species prefer estuarine habitats while others are more common in coastal waters. In the warmer, summer months, blooms or swarms of jellyfish are often a conspicuous presence in our waterways. Although jellyfish are common, little is known about the biology of many species or the role they play in the marine environment.

WHO'S WHO AND WHERE ARE THEY FOUND?

A wide variety of jellyfish are found in the waters of temperate Australia. Although many species are found in offshore, oceanic waters, we will concentrate on the nearshore species that are most likely to be encountered. The Jelly Blubber (*Catostylus mosaicus*), the Lion's Mane Jellyfish (*Cyanea capillata*), the White Spot Jellyfish (*Phyllorhiza punctata*) and the Moon Jelly (*Aurelia aurita*) are very common in nearshore waters in southern Australia. The Moon Jelly is found all around Australia (and is also widely distributed around the world), while the Jelly Blubber is only present along the east coast of the continent. The White Spot Jellyfish is sometimes abundant in New South Wales and Western Australia and the Lion's Mane Jellyfish is sometimes found in estuaries, but is more common in coastal waters extending from Western Australia to Queensland. Bluebottles (*Physalia physalis*) occur at the surface in coastal waters throughout Australia. They are frequently encountered by swimmers and have a potent sting. The Temperate Box Jellyfish (*Carybdea rastoni*) is also found close to shore, from Western Australia to New South Wales, and although it also has a painful sting, it is fortunately not as toxic as its tropical counterpart.

LIFE HISTORY OF A JELLYFISH

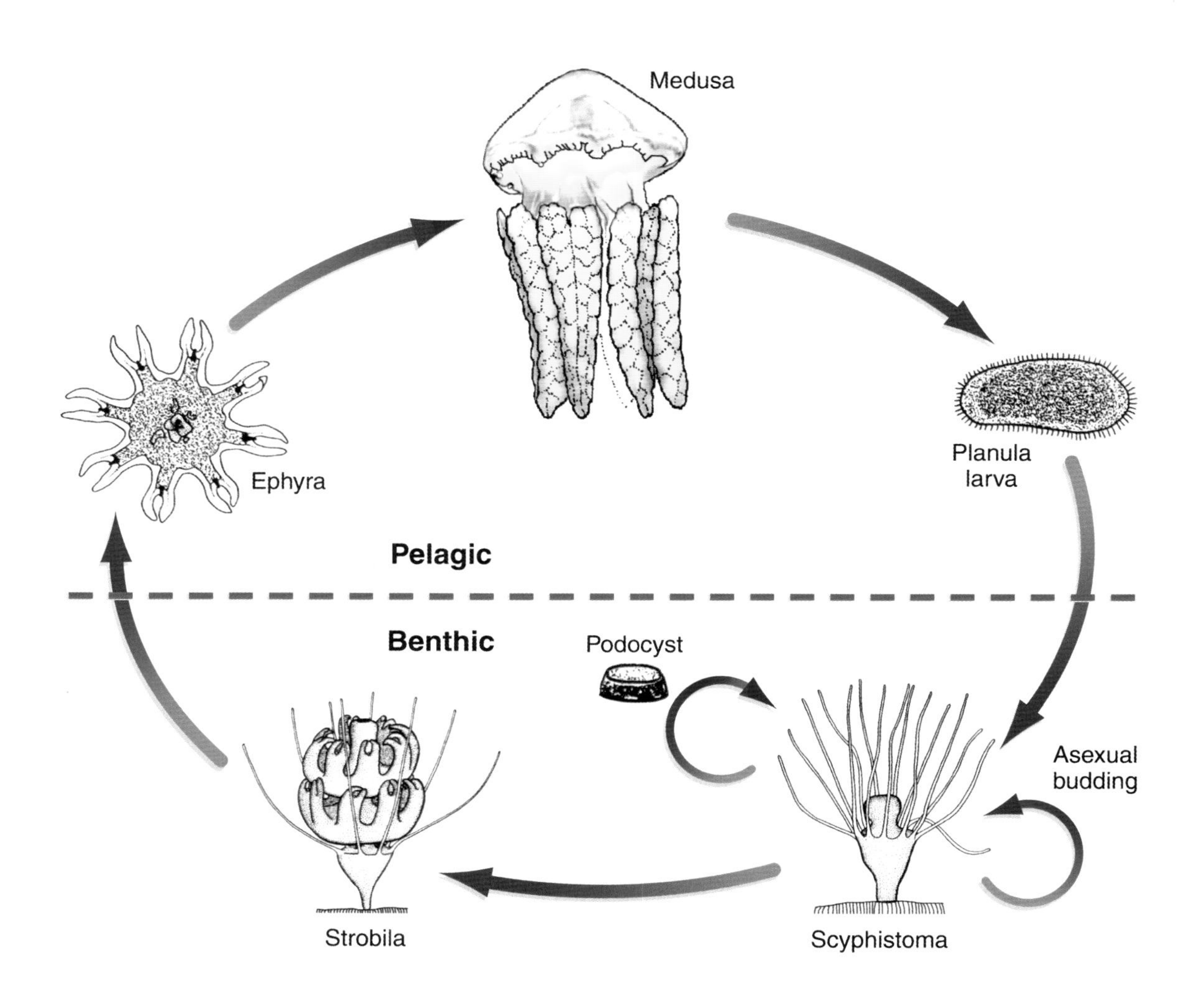

LIFE HISTORY

We normally associate 'jellyfish' with the large animals that we frequently see from boats and indeed some jellyfish can be very large. The life cycles of jellyfish are, however, complex and the large 'jellyfish' (or medusa as they are known) are really just the final stage in the development of these animals (see figure above). The typical jellyfish life cycle has separate male and female animals. In some species the females can be identified by the presence of brooding filaments on the underside of the bell (eg, Lion's Mane Jellyfish), but for many species the sexes can only be separated by examining the reproductive organs under a microscope. Mature adults release eggs and sperm which are fertilised to form larvae known as planulae. The larval stage is usually tiny, often about 0.2 mm long. In some species these larvae are brooded around the filaments or under the bell of adults, while in other species larvae may join the plankton and are free swimming. The free-swimming larvae move by beating small hairs that cover their body. The larvae do not feed and rely on stores of fats for their energy. After a few days, or weeks of development, the mature larvae attach to rocks, shells or seagrass and metamorphose into a polyp, known as a scyphistoma. The polyps are so tiny (less than 1 mm across) that divers rarely notice them. The polyps have long tentacles which, like the adults, contain stinging cells and are used to capture planktonic prey. The polyps may reproduce asexually by budding, or by passing through an encysting stage known as a podocyst. Under favourable conditions the polyps will transform directly into a strobila, and juvenile jellyfish (ephyrae) will bud from the top of the strobila. The ephyrae enter the plankton and grow into the adult medusoid stage that we commonly recognise as a jellyfish.

△
The adult medusae produce planula larvae that settle and metamorphose into polyps (scyphistomae). Polyp may undergo asexual reproduction via the formation of podocysts prior to strobilating. The ephyrae produced during strobilation join the plankton and grow into adult medusae. Modified from Calder (1982) with permission.

Most species of jellyfish are found only at certain times of the year. The scyphistomae produced one summer may survive over winter and metamorphose into the familiar jellyfish during the following summer. The resulting medusae may then only survive for a few months over the summer period. The medusae are extremely fast growing (eg, Jelly Blubbers can grow to 20 cm diameter in less than three months). Boaters and divers, therefore, can often be surprised when they find no jellyfish at a location and return a few weeks later to find the area thick with them.

JELLYFISH–FISH ASSOCIATIONS

Juvenile fish are frequently seen swimming in association with jellyfish. Often they will swim around the bell of the jellyfish but may also be found swimming amongst the tentacles. A variety of fish species have been observed doing this, such as jacks and trevallies (Family Carangidae), Jack Mackerel (*Trachurus declivis*), trevallas (Family Centrolophidae), Mosaic Leatherjackets (*Eubalichthys mosaicus*) and Yellowtail (*Trachurus novaezelandiae*). Although little is known about the relationship between the jellyfish and the fish it is believed that the fish may use the jellyfish as shelter, with the tentacles providing protection against potential predators. So how do the fish avoid being stung by the tentacles? Juvenile fish are skilful swimmers and observations of fish that swim with jellyfish suggest that the juvenile fish may actively avoid physical contact with the tentacles. When a jellyfish is approached underwater, the juvenile fish dart up beneath the bell of the jellyfish, an area where the stinging cells are less abundant. Fish may be resistant to jellyfish toxins or protected by the mucus on the fish's skin, although this phenomenon is poorly understood.

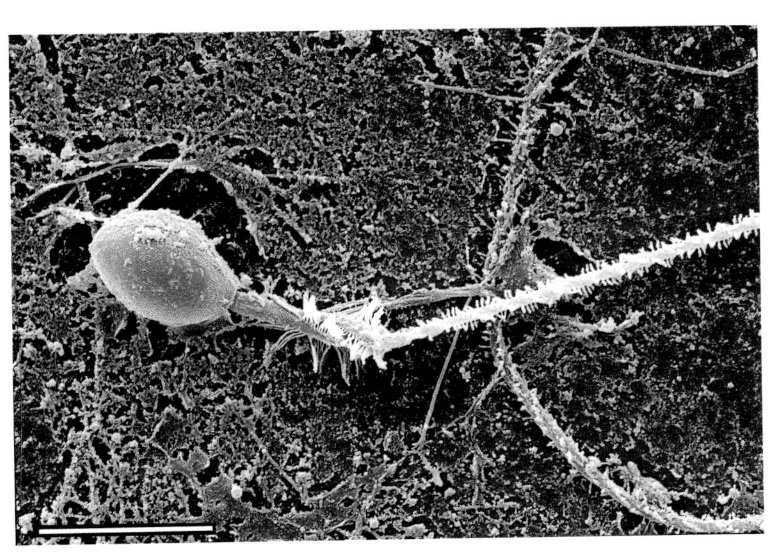

▽ Micrograph of a discharged nematocyst of the Lion's Mane Jellyfish. The discharged spear can be seen protruding from the capsule. Scale bar is 10 µm. *Robert Condon*

PREDATORS OF JELLYFISH

Although jellyfish contain little nutritional value (their main constituents are collagen and water) they are preyed upon by a surprising variety of animals. The most commonly cited predators of jellyfish are sea turtles. Turtles have hard beaks and tough, leathery mouths which makes them immune to the stings of jellyfish. Even the most toxic jellyfish, the Tropical Box Jellyfish (*Chironex fleckeri*) of northern Australia, is a favoured food of sea turtles. Turtles are, however, rare in temperate waters. It is now becoming apparent that jellyfish are also consumed by many species of fish, such as the Short Sunfish (*Mola ramsayi*), pufferfish, leatherjackets, bream, Australian salmon and some species of tuna. In fact, jellyfish tentacles are sometimes used to bait fish traps to attract leatherjackets. The biggest obstacle to identifying predators of jellyfish is the fact that jellyfish tissue is rapidly digested. Examinations of the gut contents of fish and other animals, therefore tend to overlook jellyfish and underestimate the variety of species that prey on jellyfish. Seabirds, such as gulls, shearwaters and wading shorebirds, may also prey on jellyfish when they are washed ashore. Some species of jellyfish are also cannibalistic, feeding on the young of their own species, while others may eat jellyfish of different species; for example the Lion's Mane Jellyfish will sometimes eat the Moon Jelly. The polyp stage of the life cycle may also be consumed by predators such as crabs, amphipods, sea spiders (pycnogonids) and sea slugs. Like adults, polyps may also consume each other.

WHAT DO JELLYFISH FEED ON?

Jellyfish feed by filtering small animals out of the water. They primarily feed on plankton, such as fish eggs, larval fish and small crustaceans including crab and lobster larvae and copepods. Some species, however, feed on larger, juvenile fish. Although intuitively we would expect jellyfish feeding behaviour to be passive, jellyfish have evolved mechanisms that increase the chances of them encountering prey. Some species actively move to areas or depths where currents converge or where water of different temperatures or salinities meet. These areas are usually rich in plankton, and jellyfish are more likely to encounter prey there. It has also been suggested that jellyfish may use chemical cues to detect areas abundant in plankton. Most species swim and feed by rhythmic contractions of their bell which generate water currents that bring plankton

into close proximity with the tentacles. Once organisms encounter the tentacles, the stinging cells fire and trap the prey item. Depending on the type of jellyfish, the tentacles either pass the prey directly to the mouth or the food items pass along a duct that runs along the length of the tentacle and opens into the stomach cavity, in the bell of the jellyfish. Food is digested within the stomach cavity and waste products are then expelled back through the mouth. Dense aggregations of jellyfish may severely deplete the abundance of plankton and in the northern hemisphere the low abundance of some commercially important fish species has been attributed to intense predation on fish larvae by jellyfish.

In addition to gaining energy by feeding on plankton and small fish, some jellyfish species have the ability to photosynthesise. Like corals, some jellyfish carry tiny plant cells (zooxanthellae) embedded in their tissue. These cells use energy from sunlight to produce organic compounds which are then utilised by the jellyfish. This unusual relation-

◁ An unidentified species of large jellyfish seen in oceanic waters of Jervis Bay, New South Wales. *Kevin Deacon*

ship between the jellyfish and zooxanthellae is beneficial for both organisms as the zooxanthellae provide nutrition for the jellyfish and in return the jellyfish provides a suitable habitat for the zooxanthellae.

WHAT CAUSES JELLYFISH TO STING?

Jellyfish are notorious for the painful sting they can inflict. The sting is primarily used in prey capture, but is also an effective defence mechanism. Stings are inflicted in response to chemical and mechanical stimuli. So what causes the sting? Jellyfish contain an abundance of microscopic capsules called nematocysts. Nematocysts are found primarily on the tentacles, but may also be present on other parts of the animal or released in mucus that is exuded from the jellyfish. These complex capsules contain a barbed, spear-like structure which is discharged when stimulated. Undischarged nematocysts lie inverted within the capsule. When triggered, the nematocyst is propelled from the capsule, and turns inside out to expose the barbs which help to anchor the spear in place. The entire discharging process may take place in as little as 3 milliseconds and the rapid acceleration of the spear enables the nematocyst to penetrate structures as tough as the shells of crabs. A protein toxin passes along the length of the spear and penetrates the body.

The reaction of people to jellyfish stings varies depending upon the species encountered and which part of the body is stung. The soles of the feet and the pads of the fingers have thick layers of keratin which provide some protection, however stings to other less protected parts of the body, such as the back of the hand, produce a more severe reaction. In general, localised pain and skin irritation occurs and in more severe cases people may experience nausea and muscle cramps. Treatment of jellyfish stings is varied and in serious cases medical help should be sought. The first step should be to remove any remaining jellyfish tentacles to prevent further nematocyst discharge. Nematocyst discharge of box jellyfish can also be inhibited by the application of vinegar. However vinegar is not recommended for treatment of other jellyfish stings and the reader is referred to Williamson et al. 1996 in the Further Reading section for this chapter for the medical treatment of jellyfish stings.

▽ Lion's Mane Jellyfish. Portsea, Victoria. *Rudie Kuiter*

◁
The jellyfish *Pseudorhiza haeckeli* is common in subtropical regions. New South Wales. *Ken Hoppen*

HOW DO JELLYFISH MOVE?

Jellyfish use a variety of mechanisms to move. Some species move with the wind or currents, such as the Bluebottle, while others actively swim. Although not as fast as most fish, some species of jellyfish are relatively strong swimmers. Large specimens of the Jelly Blubber have been observed swimming horizontally at a rate of 8 m/min. Some species of jellyfish therefore have some ability to control where they move and are not completely at the mercy of currents. The rhythmic contraction of bands of muscle that run around the margin of the bell cause the bell to pulsate and water is pushed out behind, propelling the jellyfish forward. Although smaller species, such as the Moon Jelly, use a similar mechanism to move, their small size suggests that they are less effective swimmers and probably more vulnerable to the movement of currents. However, any species which can actively swim may be able to alter their vertical position in the water column. Currents often move in different directions or speeds at different depths and jellyfish may be able to use this to control where they move.

Other species have adopted more novel means of locomotion. The common Bluebottle has a gas-filled sac, or float, which suspends the Bluebottle at the water's surface. The float is caught by the wind and blows the Bluebottle around the ocean. Bluebottles can be a common and irritating presence on summer beaches. A curiosity in the morphology of the Bluebottle is that some individuals have floats that are orientated 45° to the right of the wind, while others have floats orientated 45° to the left of the wind. This means that jellyfish disperse in different directions, a by-product of which is they are more evenly dispersed and may compete less intensely for food. Similarly the By–The–Wind–Sailor (*Velella velella*) has a sail-like structure on the top of the bell which also catches the wind.

JELLYFISH FISHERIES

Jellyfish are commonly eaten in Asian countries. They have been collected off the coast of China for more than 1000 years and large-scale commercial harvesting of jellyfish has occurred in Asia since at least the 1950s. During 1992 more than 350 000 t of jellyfish were harvested worldwide. Typically, the bell of the jellyfish is dried using a combination of alum and salt to produce an elastic, rubbery product. Jellyfish is prepared as an ingredient for cooking by rehydrating it in fresh water before cooking. Although it is not very well known, a fishery for jellyfish was established in New South Wales in 1996 with a total allowable catch of 1500 t biannually. Plans are now underway to develop jellyfish fisheries in Victoria, and possibly Queensland and Tasmania.

Small juvenile Southern Rock Lobsters are found hidden in small holes, with only their banded antennae protruding. Watsons Bay, Tasmania. *Matt Edmunds*

12 Southern Rock Lobsters

Stewart Frusher, Jim Prescott and Matt Edmunds

Southern Rock Lobsters (*Jasus edwardsii*) are among the largest animals on temperate reefs. They live in a wide range of habitats and depths, from sheltered embayments to exposed coasts, and from shallow rock pools to deep reefs on the edge of the continental shelf. Southern Rock Lobsters are the basis for important commercial fisheries in South Australia, Victoria and Tasmania.

Southern Rock Lobsters are easily distinguished from the Western Rock Lobster (*Panulirus cygnus*) by their shorter antennae and sculptured shell surface. Seven *Jasus* species occur in the temperate waters of the southern hemisphere, two in Australia (the second species is the Eastern Rock Lobster, *J. verreauxi*; see following chapter). Southern Rock Lobsters are found along much of Australia's southern coastline — from Dongara, Western Australia, to Coffs Harbour, New South Wales — but the species' commercial distribution ranges only from the Great Australian Bight to southeastern Victoria and Tasmania. Southern Rock Lobsters are also abundant along the coast of New Zealand.

The Southern Rock Lobster was originally thought to be the same species as the South African Lobster (*Jasus lalandii*), which was first described in 1837. Since then its taxonomic status has undergone numerous changes. In 1963, Southern Rock Lobsters in Australia were thought to be a separate species and given the name *Jasus novaehollandiae.* At the same time, the New Zealand population was renamed *Jasus edwardsii.* Modern genetic methods could not separate the Australian and New Zealand species, so the name *Jasus edwardsii* is now applied to both populations. The genetic similarity of New Zealand and Australian populations demonstrates there is mixing of the populations. This mixing occurs by larvae crossing the Tasman Sea, settling and reproducing with the local population.

Southern Rock Lobsters inhabit crevices in rocky reefs from the intertidal zone to the depths of 200 m. The lobsters are fished throughout this depth range using baited traps (termed 'pots' by the fishers). Occasionally, fishers catch Southern Rock Lobsters at 300 m. Although lobster numbers have been greatly reduced by 50 years or more of commercial and recreational fishing, they are still abundant and support the highest value inshore commercial fisheries in South Australia, Victoria, and Tasmania. Southern Rock Lobsters also support important recreational and customary fisheries throughout their range.

To the untrained eye the sex of a lobster may not be obvious; however, four clues help identify the sexes. Fishers and scientists most commonly use the presence of double swimmerets or pleopods under the female's tail. The pleopods of female lobsters are lobed, with the two lobes having different functions. The inner lobe is a rod-like structure that, once the female is sexually mature, bears long hairs to which the eggs attach. The presense of these setae can be used as an indication of sexual maturity when the females are not carrying eggs. The outer lobes are flat, leaf-like appendages which help protect the eggs and maintain water circulation, providing oxygen to the developing larvae. Males have smaller non-overlapping pleopods without setae. Females can also be identified by small pores at the base of the third pair of legs, from which the eggs are extruded, while males' external openings for the sex organs are at the base of the fifth pair of legs. Females' fifth pair of legs terminate in a small claw or 'nipper', unlike males where the tip of each leg ends in a simple point. Females use the claw to tend the eggs, removing any foreign matter and non-fertile eggs. Finally, females tend to have broader tails than males to facilitate carrying large broods of eggs, while mature males tend to develop massive carapaces and legs which play a role in courtship and mating.

REPRODUCTION AND EARLY LIFE

Southern Rock Lobsters have a remarkable life cycle in which they change appearance as they grow through several distinct phases. Adult lobsters mate in autumn. Male lobsters compete for spawning females, and larger

▽ A Southern Rock Lobster sheltering under a rock ledge. Waterfall Bay, Tasmania. *Simon Talbot*

males attract and mate with more females than smaller males. Males held in aquaria are capable of successfully mating with up to 16 females. During mating, males deposit a gelatinous sperm package on the female's underside, between her walking legs. The sperm packet is broken open to fertilise the eggs. The female extrudes her eggs which attach to the pleopods under the tail. The number of eggs each female bears depends on her size. Larger females carry more than 600 000 eggs. Recently extruded eggs of lobsters in shallow water are a dark burgundy red while those from lobsters in deeper water are lighter orange. As the eggs develop they become lighter in colour, revealing the developing larvae inside. Careful examination of late-stage eggs reveals small black spots — the eyes of the larvae. The Southern Rock Lobster spawns only one brood of eggs/year, unlike the Western Rock Lobster and other tropical Australian species. Spawning females are found from shallow inshore waters to deep waters on the edge of the continental shelf. During hatching, females congregate on the deeper edges of the reefs they normally inhabit.

The eggs hatch after 3–5 months incubation over winter and spring and the tiny (2 mm) phyllosoma larvae disperse into the water column as plankton. They are soon carried offshore by the currents where they enter the pelagic ecosystem. The planktonic phyllosoma larvae have a flattened, transparent body similar to that of the Eastern Rock Lobster (see photo in following chapter). The phyllosoma can swim and capture smaller planktonic prey. However, they are not strong swimmers and are largely at the mercy of the currents. Southern Rock Lobster phyllosomas, like their western relatives, make vertical migrations, rising near the surface at night. Phyllosomas may have some influence over where they go by migrating vertically into currents of different directions. If they survive their larval phase they may have travelled many thousands of kilometres, trapped in giant ocean eddies or in one of the currents flowing past southern Australia. Off New Zealand, Southern Rock Lobster larvae have been found at highest densities as far as 1200 km from the coast. Oceanic life may last for as long as two years, but most larvae spend about 16–23 months in the plankton.

The phyllosoma larvae go through 11 morphologically distinct stages, growing from about 2–50 mm total length (some giants of 70 mm have been found). The final-stage phyllosoma is ready to metamorphose into a puerulus larva if, and only if, it receives the right environmental cue. Metamorphosis takes place in oceanic waters adjacent to the continental shelf. The metamorphosis from phyllosoma to puerulus is dramatic: the larva's total length decreases by 50 per cent; the carapace is radically reduced in width; and the abdomen (tail) grows substantially and becomes a powerful means of swimming. The puerulus resembles a miniature adult (carapace length [CL] of about 11 mm and total length of approximately 25 mm), but has few spines and is transparent. After metamorphosis, pueruli swim across the continental shelf (as much as 60 nautical miles) and settle on coastal reefs.

Although it appears that the metamorphosis of phyllosoma to puerulus is triggered by the proximity of land, the precise cues for metamorphosis are unknown. Likewise, it is uncertain what cues are used by the puerulus to find an inshore reef and finally choose a suitable site to settle and begin its benthic life. The puerulus stage is thought to be a non-feeding stage with the sole purpose of bridging the gap between the oceanic phyllosoma phase and the benthic-dwelling lobster. The puerulus has an extra pleopod or swimmeret attached to the end of each pleopod beneath the tail. This 'double paddle' aids the puerulus in swimming and is shed on moulting after the puerulus has settled onto coastal reefs. After settling, and prior to the moult to the juvenile, the transparent puerulus develops the adult pigmentation initially around the antennae, followed by the carapace and then the tail.

With such an extensive larval life, there is potential for considerable mixing of larvae from the entire distribution of the species: settlement in one region may have no relationship with egg production in that region. The amount of settlement in any region would also be influenced by oceanic currents, food supplies and the success of potential predators. The numbers of pueruli settling on reefs provides the first indication of future recruitment to that region. Recently begun puerulus settlement monitoring programs for Southern Rock Lobsters are important, both for managing sustainability of the resource and in predicting future catches. Puerulus monitoring projects take advantage of the cryptic nature of puerulus and use sloping wooden plates which form wedge-shaped crevices to create an artificial habitat. Several plates are bolted together in a standardised design called a crevice collector. These larvae collectors are positioned on the seabed in clusters along the coastlines of both New Zealand and Australia. Collectors are serviced by marine biologists at monthly intervals in many locations throughout southern Australia.

Ocean currents play an important role in facilitating puerulus settlement on coastal reefs. Similarities in settlement are seen around a number of sites in South Australia, Victoria and northern Tasmania. Although settlement has been monitored for less than ten years, settlement was poor during the 1996–8 El Niño period in most locations. The dominant ocean currents which affect these sites are southern extensions of the Leeuwin Current in South Australia, western Bass Strait and northwestern Tasmania, and the East Australian Current in eastern Bass Strait and northeastern and eastern Tas-mania. Both currents are subtropical in origin and poor in nutrients. In contrast, puerulus monitoring sites in southeastern Tasmania are bathed by the cooler waters of the subantarctic convergence which brings nutrient-rich waters to southern Australia, and showed increased puerulus settlement during the recent El Niño.

Pueruli settle throughout the year, but show strong seasonal peaks in settlement. In northern Tasmania, Victoria and South Australia, settlement occurs mainly during winter and less in summer. In southern Tasmania, the opposite trend occurs. While there is consistency in the timing of peak settlement periods, differences there may be substantial differences in the magnitude of settlement between sites less than a kilometre apart.

Pueruli have been found on anthropogenic structures recently installed in shallow-water regions such as piers, oyster racks, mussel lines, mooring lines and fouled boat hulls. Whether such structures attract puerulus from other habitats or whether they facilitate increased settlement is unknown. Recent investigations attached strips of trawl mesh to puerulus collectors in an attempt to simulate the three-dimensional structure of kelp on reefs. The additional structural complexity or surface area enhanced puerulus settlement. This work indicates that effects on habitat, such as the large-scale loss of *Macrocystis* kelp beds in Tasmania and elsewhere have the potential for significant impacts on lobster recruitment.

LIFE AS A JUVENILE

Adjacent areas of reef can have substantially different population densities of juveniles; settlement on offshore reef habitats has not been studied, so we are uncertain how much settlement occurs there and whether these areas depend on migrations of juveniles from inshore reefs to maintain the local population. Tagging studies discussed later have been restricted to lobsters caught from pots, so have not included many lobsters less than 60 mm CL. Divers have reported seeing large aggregations of small lobsters (30–50 mm CL) in inshore waters; these may be aggregating prior to dispersing or migrating to deeper water reefs, similar to the migration of juvenile Western Rock Lobsters.

Pueruli and small juveniles lead a solitary existence. These lobsters begin their life on the reef with a mottled brown and cream camouflage colouration, which tends toward a more uniform red colour as they grow. Early benthic-phase juveniles have a habitat preference for individual holes not much bigger than themselves. In southern Australia, reef substrata are primarily hard igneous rocks (such as granite or basalt), or softer sedimentary rocks (such as sandstone and limestone). The sedimentary rocks tend to weather more rapidly and provide a greater number of small holes for early juveniles, compared to the harder and less eroded rocks such as granite. Lobster abundance on some reefs composed of hard rock may be limited by shelter availability for the early juvenile stages. Suitable shelters for early benthic-phase lobsters is critical, because their small size makes them vulnerable to many predators. Most of the common reef fishes, including wrasses and leatherjackets, make an easy meal of juvenile lobsters outside their shelters. Large crabs and octopuses also pose a major threat. Consequently, small lobsters remain well hidden in their holes, only their relatively long, banded antennae poking out. Small juveniles eat highly cryptic animals such as brittle stars, small crustaceans and bivalves. These prey are normally associated with kelp holdfasts, encrusting algae and small crevices in the reef.

From approximately 35 mm CL and larger, juveniles become more gregarious, sharing crevices with other juveniles and adults. At this gregarious stage of their life history, juvenile lobsters tend to prey less on cryptic organisms and forage more widely for prey such as sea urchins, chitons, snails and crabs.

△
Den of juvenile and adult Southern Rock Lobsters. Forestier Peninsula, Tasmania.
Matt Edmunds

LIFE AS AN ADULT LOBSTER

Female lobsters mature about five years after settling onto the reef. The size at which females become sexually mature depends on their growth rate, which in turn depends on where they were during their adolescence. Remarkably, females only 10 km apart can mature at different sizes.

Observations of dense lobster populations in Marine Protected Areas and the catch rates reported by pioneer fishers suggest that lobsters were once extremely abundant on temperate reefs. There is little doubt that lobsters were important predators of molluscs and sea urchins, the numbers of which in turn would have affected other aspects of the reef ecosystem, such as algal cover. Numerous studies have examined the relationships among prey organisms (see Chapter 15), but none has studied directly the effect of lobster abundance on prey species. Consequently, we can only speculate about the effect reductions in lobster biomass have had on the temperate reef ecosystem. In many regions, the size of the harvestable portion of the population is thought to have been reduced by more than 90 per cent.

GROWTH

Southern Rock Lobsters, like all crustaceans, grow by periodically shedding their complete exoskeleton in a process known as moulting. Close scrutiny of a lobster's discarded shell reveals that all surfaces in contact with an external medium are shed. This includes the lens of the eye, all the fine hairs covering the body, the external surface of the gills and mouthparts and the lining of the digestive tract including stomach and intestine. After shedding the old exoskeleton, lobsters increase their body size by increasing the water content of the cells in their body. Once the new shell has hardened at the enlarged size, the lobster replaces the water with tissue during the period until the next moult. The number of moults which a lobster undertakes is dependent on its size and maturity. In general, juvenile lobsters moult several times each year and the number of moults/year decreases as they approach sexual maturity. At sexual maturity, energy reserves which were partitioned between metabolism and growth become partitioned between metabolism growth and gonad development. After reaching sexual maturity, lobsters normally moult only annually and female growth is often reduced because egg development places a high demand on energy reserves. Mature females moult primarily between April and June and mature males from September to November. The timing of moulting can vary from year to year depending on environmental factors, larger individuals moult earlier than smaller individuals.

In addition to reproductive development, energy reserves in lobsters are also used to regenerate damaged body parts, such as lost limbs and minor tail and body damage. Regeneration of damaged or lost body parts has higher priority for using energy reserves than either growth or reproduction. Lobsters in the process of regenerating limbs grow at a slower rate and damaged females tend to have smaller egg clutches. Severely damaged females may not produce a brood at all.

Growth rates vary considerably in the Southern Rock Lobster because of differences in temperature (the slowest growth rates are found in deeper waters off southern Tasmania) and density-dependent factors, such as food or shelter. The fastest growth rates are found around the Bass Strait islands, particularly King Island, and near the head of the Great Australian Bight. Within regions, growth rates decrease with increasing depth, probably because shallower waters are warmer and have increased productivity. Males grow more rapidly than females for much of their life and reach greater weights in unfished populations. Large males can reach over 220 mm CL and weigh more than 6 kg, while females can reach 175 mm CL and weigh 4 kg.

In certain regions, such as the Coorong region in South Australia and in southern Tasmanian waters, the density of lobsters is high and lobsters have lower growth rates. The average catch rate of lobsters in southern Tasmania can exceed 50/pot (95 per cent undersized) whereas less than 5 lobsters/pot is characteristic of most other areas. The high density of lobsters is partly caused by the slow growth rates which have been recorded in tagging studies. Several female lobsters tagged on reefs in shallow water in southern Tasmania in 1976 were recaptured in 1994 and were still undersized. Most of these females had grown less than 2 mm CL/year and two individuals grew less than 1 mm CL annually. The lobsters

▽ Lobsters can live in dense aggregations within shelters. The largest lobster will force the smaller ones to the front if a predator approaches. Forestier Peninsula, Tasmania. *Matt Edmunds*

caught from the 1976 tagging event were 74–104 mm CL (only part of the size range) and provide a good demonstration of their potential longevity.

Adult lobsters are omnivorous and feed on a wide variety of organisms: crabs and other crustaceans, sea urchins, molluscs including bivalves, chitons and gastropods, and a variety of algal species. Diet and food availability influence the growth and ecology of lobsters, and also affect the colour of the exoskeleton and eggs. Lobsters in shallow water are generally dark red in colour; this fades to a strawberry colour in depths around 40 m and then to whitish pink in even deeper water. In deep water, often only the ridges on the carapace and the dorsal region of the lobster's tail are light red and the remainder of the carapace and the sides of the tail are cream coloured. This change in pigmentation is also reflected in the eggs. Light-coloured lobsters from deep water return to the darker red colour during subsequent moults when released in shallow water. This change in colour is associated with diet, as lobsters held in aquariums and fed on fish diets become lighter with each moult. A group of mature females held over two spawning seasons also produced cream-coloured eggs in the second season after being fed on a fish diet. Lobsters are unable to synthesise the reddish carotenoid pigments which determine the colour of their shell and eggs. This must be obtained from their diet, either directly from the algae which produce the carotenoids or indirectly from animals which have fed on the algae. Because of their reliance on sunlight, algae are most common in shallow water.

MOVEMENT

In southern Australia, no annual or regular large-scale movement or migration patterns have been found. Some tagged lobsters have moved up to 114 km, but these are the exception rather than the rule. Most tagged lobsters have remained within several kilometres of the position where they were tagged and released. In Tasmania, most movement occurs in the north of the state where growth rates are fastest and lobsters are younger for a given size. Movement may therefore be associated with adolescent behaviour. This could explain the lack of movement in southern waters where growth rates are slow and thus few adolescent animals have been captured and tagged. In South Australia there are two distinct areas from which lobsters migrate long distances. These are reefs at the southern end of the Yorke Peninsula and the Coorong, both of which are atypical lobster habitats. Long-distance movements in Victorian waters have been observed only in waters east of Cape Otway.

Documented movement is generally offshore or along reef margins parallel with the coast. Offshore movements are reported by fishers who routinely find red lobsters in deep water. There have been few records of lobsters moving inshore. On a smaller scale, lobsters move out of their dens during the mating period. New Zealand researchers have demonstrated that there is a size-related hierarchy during mating, with larger male lobsters being dominant. Excursions from reefs onto adjacent sand flats have been observed in New Zealand and Tasmania, the latter found during trial trawling operations.

The commercial importance of Southern Rock Lobster has resulted in considerable research on its biology. However, much of this research has been directed towards aspects applicable to management of the fishery. There is still much to be learnt about this intriguing animal, particularly about the ecology of post-puerulus and juveniles stages, and interactions between this species and other ecologically important species living on temperate reefs.

A juvenile Eastern Rock Lobster. Bermagui. New South Wales. *Rudie Kuiter*

13 Eastern Rock Lobsters

Steven Montgomery

The most abundant of several species of rock lobster in New South Wales is the Eastern Rock Lobster (*Jasus verreauxi*), the largest rock lobster in the world. Eastern Rock Lobsters are also found in Victoria and occasionally Tasmania, and along the South Australian coast as far as Port MacDonnell. The species is also found in New Zealand, predominantly around the North Island. Stocks in New Zealand and Australia are genetically distinct, suggesting little exchange of larvae between the two countries. Eastern Rock Lobsters are fished both commercially and by recreational divers and trappers. The species accounts for more than 97 per cent of the rock lobsters caught in New South Wales, the remainder being mostly Southern Rock Lobsters (*Jasus edwardsii*). In this chapter I describe what is presently known about the ecology of Eastern Rock Lobsters. Because research on this species has focused on the commercial fishery, little attention has been paid to the biology and ecology of the species. The information in this chapter comes from data collected by scientists, fishers and other keen observers of this species.

REPRODUCTION AND EARLY LIFE

Female Eastern Rock Lobsters become sexually mature when they are around 166 mm long (all lengths given are carapace lengths) but females in the north of the state may mature at larger sizes than in the south. Females spawn between September and January each year and carry only one brood of eggs/season. The mechanics of spawning and fertilisation of the eggs are similar to those described for Southern Rock Lobsters in the preceding chapter. Eastern Rock Lobsters are highly fecund. For example, large females of the same species in New Zealand carry around 1.9 million eggs, many more than have been reported to be carried at any one time by females of other species of rock lobster. Most female Eastern Rock Lobsters carrying eggs are found in waters shallower than 100 m.

The life cycle of the Eastern Rock Lobster follows the same complex series of forms and habitats as the Southern Rock Lobsters. The larvae are thought to spend 8–12 months in the plankton on a journey that may take them hundreds of kilometres offshore and from the surface to depths greater than 400 m. As is the case for Western Rock Lobsters, the behaviour of phyllosoma larvae may place them in currents that transport them offshore in surface waters during development and subsequently return them to the coast as pueruli in deeper currents. Although pueruli are probably assisted back to the coast by currents, they can also swim quite strongly.

Metamorphosis from phyllosoma to puerulus is thought to occur near the edge of the continental shelf. Many commercial fishers have found pueruli of Eastern Rock Lobsters amongst their traps set at various depths across the shelf. The puerulus stage lasts up to a month, during which time it changes from translucent to a dark colour more suited to life on a rocky reef. In common with other rock lobsters, pueruli of Eastern Rock Lobsters do not feed and grow little through this stage of their lives because their mouthparts are not completely formed.

In New South Wales, pueruli settle onto shallow rocky reefs between August and January each year. This single, consistent season of settlement is different from that of the Southern Rock Lobster, which has settlement on nearshore reefs throughout the year with one or more peaks in abundance. As with other rock lobster species, however, there are large differences among years in the numbers of pueruli settling. Similarly, there are differences in the density of pueruli among locations along the coast. More pueruli settle in waters south of Newcastle than to the north. This pattern in the relative abundance of pueruli is reflected in the abundance of juveniles.

Regional differences in puerulus abundance may be partially explained simply by the width of the continental shelf and the strength of the East Australian Current (EAC, see Chapter 1). Larvae from lobsters that spawned in waters in the north of New South Wales are probably transported long distances south by the EAC to areas where the continental shelf is narrower and the current weaker. If so, the chances of pueruli reaching the shore may be greater in the south. Further north, the EAC might carry them far from shore as it breaks away into the Tasman Sea. Other possibly related explanations include the impact of different water temperatures on the survival of larvae, different rates of predation, and patterns of movement during the phyllosoma stage.

△ Pueruli of the Eastern Rock Lobster change colour from transparent to brown during their development. *Steven Montgomery*

LIFE ON NEARSHORE REEFS

Settled pueruli and small juvenile Eastern Rock Lobsters have not been observed in the wild. Their ecology may be similar to that of other species of rock lobster. If so, they live solitary existences amongst the complex structure provided by large brown algae such as Crayweed (*Phyllospora comosa*) and the Common Kelp (*Ecklonia radiata*), in seagrass beds and in rock pools in the intertidal zone. Juvenile Eastern Rock Lobsters spend approximately three years of their lives on nearshore rocky reefs, during which time they grow to be up to 130 mm long and are most abundant on rocky reefs in the south of the state. On any particular reef, smaller lobsters are found in shallower water than larger ones. During this period they become increasingly gregarious and are often seen in aggregations thereafter, probably to enhance protection from predators and to facilitate their complex and poorly understood social and mating behaviour.

The diet of lobsters during this phase of their lives includes worms, small bivalves, echinoderms and fish, whilst the main predators are fish, sharks, rays and octopuses. As the lobsters become larger, they may in turn have an impact on the abundance of such important species as sea urchins and abalone. The feeding behaviour of these nocturnal foragers, and their effects on reef communities, are productive areas for study.

When they are approximately 120–130 mm long, Eastern Rock Lobsters move to reefs in water deeper than 20 m and are found at the edge of the continental shelf in depths to 200 m. Their habitat consists of overhangs and caves among rocky reefs but also includes soft sediment. Commercial fishers think lobsters bury in soft substrate. Most of these small lobsters in the south of New South Wales are still sexually immature; mature lobsters are mostly found north of Newcastle. This geographical pattern in the sizes of lobsters is atypical among species of rock lobster. The causes of this apparent difference in sizes of lobsters with latitude is unknown but three hypotheses are consistent with the known patterns: (i) differences in growth rate with latitude, (ii) northward migration or (iii) differences in behaviour toward traps. Generally, mature rock lobsters are distributed throughout the geographical range of the species. Geographic separation of spawning-size Eastern Rock Lobsters is also found in New Zealand.

Mature lobsters are found in greatest numbers on reefs shallower than 100 m over spring and summer and on reefs in deeper water over autumn and winter. The occurrence of mature Eastern Rock Lobsters in shallower water appears to be associated with spawning — most mature females are carrying eggs at this time. Aggregation in deeper water over autumn and winter may be associated with moulting and mating. These hypotheses have yet to be tested, but similar patterns in reproduction and associated movements have been described in other species of rock lobster.

▽ Eastern Rock Lobsters are the largest rock lobsters in the world; the lobster on the left is longer than the minimum size at which they can be harvested. *Steven Montgomery*

MIGRATIONS

Long-distance migrations are a feature of many species of rock lobster. Evidence for such migrations in Eastern Rock Lobsters is patchy, but patterns in the sizes and numbers of lobsters along the New South Wales coast are consistent with northward migrations. The same species in New Zealand migrates long distances, probably to spawn in northern waters. If the hypothesis of a northward spawning migration in New South Wales is correct, then the main nursery grounds would be in the south of the state and the main spawning grounds would be in the north. Lobsters on southern reefs would migrate north on a one-way trip to spawn. The movement of those starting the journey further north may be more complex and involve an inshore–offshore movement. In addition to northward movements of juvenile lobsters, mature lobsters are also thought to move offshore during autumn/winter.

Under this hypothesis, larvae from the main spawning areas in the north of New South Wales are dispersed by the EAC southwards to areas where the width of the continental shelf is narrowest and where the strength of the EAC is weakest as the current breaks up into eddies. Consequently, larval Eastern Rock Lobsters would be distributed across many types of habitat by the East Australian Current. Variations in oceanographic processes would cause substantial variation in the survival of larvae and post-larvae but these links are poorly described in most species of rock lobster.

THE NEW SOUTH WALES FISHERY

The fishery for Eastern Rock Lobsters is important locally, but small on an international scale. Before the introduction of individual catch quotas in 1993–4, average annual landings by the commercial fishery were around 122 t and probably have not been much greater than 300 t/year over the history of the commercial fishery. The current minimum size limit of 104 mm means that most female Eastern Rock Lobsters caught are sexually immature. Preliminary surveys of recreational fishers suggest that the catch by the recreational fishery may be as much as 20 per cent of that caught by the commercial fishery. The commercial fishery operates almost exclusively by trapping, whilst the recreational fishery uses both trapping and gathering by hand. Amongst other regulations, individual catch quotas and possession limits are imposed on commercial and recreational fishers respectively.

A Western Rock Lobster on reef. Mandurah, Western Australia. *Sue Morrison*

14 Western Rock Lobsters

Bruce Phillips and Roy Melville-Smith

Rock lobsters are found all around the coast of Australia but the Western Rock Lobster (*Panulirus cygnus*) is only found in Western Australia. It is the most numerous and supports the most valuable single-species fishery in the nation. Because of its commercial importance, the Western Rock Lobster and its environment have been extensively studied. The species occurs mainly between Cape Leeuwin and Shark Bay, but small numbers can be found as far north as Exmouth and around the south coast to Windy Harbour.

The usual view obtained by divers is of lobsters sheltering together during the day in groups of ten or fewer animals in caves and crevices in the shallow coastal limestone reefs. This communal sheltering behaviour confronts would-be predators, such as fish, sharks and rays, with a mass of waving, spiny antennae, which are likely to be far more formidable to the intruder than just a single pair of antennae. At night lobsters move out of their caves and crevices to forage for food, particularly in seagrass beds. Juvenile lobsters eat mainly molluscs, but also crustaceans, worms, sea urchins, fish and other small animals, as well as red and green algae that occur as epiphytes on stems of seagrasses. Their eating habits have led to those less concerned with product marketing subtleties terming them 'cockroaches of the sea'.

Although lobsters grow continually, they show visible increases in size only when they moult. Before each moult, they form a soft new shell underneath their existing hardened shell. During the moult, the lobster slowly extracts its legs, antennae and tail from the old shell and finally flips out backwards. The whole process takes only a few minutes and in the act, the lobster leaves behind the old shell, together with the old gut lining, outer layer of the gills and outer lens of each eye. The animal inflates its new shell while it is still soft by ingesting and absorbing seawater. For several

days before and after each moult the lobsters are incapable of feeding because their mouthparts are soft. During this period of vulnerability they shelter in their dens.

In their first two years, juveniles moult at frequent intervals through the year depending on food availability and water temperature. As they grow, the frequency of their moulting slows down and by their fourth year they generally moult only twice/year. No one is sure of the potential lifespan of the Western Rock Lobster; however, a specimen caught when newly settled and kept alive in aquarium facilities at the Fisheries Western Australia, Marine Research Laboratories, died in 1996 aged 28. Although considerably bigger lobsters than the aquarium-held specimen have been captured in the wild, there is no way of knowing whether they were older or had simply grown faster.

When juveniles living in the shallow coastal reefs reach 4–5 years of age they migrate offshore onto the continental shelf into depths of 30–200 m. Lobsters on these migrations can move quite quickly and have been recorded moving at least 5 km/day, on occasions for several weeks on end. The migration usually starts about mid-November each year and lasts until about the end of January. The animals that make these offshore migrations tend to be lighter in colour than the rest of the population and are called 'whites' by the fishers, as opposed to the rest which are called 'reds'. Many of the migrating whites are legal size and are a major target of fishers.

In addition to the offshore migration, some white lobsters migrate along the coast on the edge of the continental shelf. These alongshore migrating lobsters tend to move northward and some have been recorded to move more than 400 km from where they were tagged. Most lobsters undergo only a single migratory moult, but a small percentage undergo a second 'whites' moult a year later and participate in a second migration taking them even further northward along the continental shelf.

In the northern part of the range of the Western Rock Lobster, the commercial fishery targets a deep-water migration which takes place each February, after the more widespread coastal migration. This 'Big Bank Run' comprises mostly white lobsters moving north in depths of 140–220 m. Recovery of tagged lobsters indicates that the run comes from both the Houtman Abrolhos Islands and coastal

△
Juvenile Western Rock Lobsters sharing a den on an inshore limestone reef. West Australia. *Clay Bryce*

areas, and some may have migrated from as far south as Dongara. Although this mass of migrating lobsters has yet to be viewed underwater, they are caught by fishers in a narrow band that must resemble a moving stream along the seabed.

In some years, lobsters in this run show obvious evidence of having migrated over a great distance, such as worn body parts, lack of energy when handled and reduced weight for their size. Mortality rates during this run must be high in those years. The number and condition of the lobsters taking part in the Big Bank migration may be related to the strength of the white's year class recruiting to the fishery in a particular season. When large numbers of whites appear on the coast, many of these lobsters participate in the Big Bank migration. In some of these seasons of strong migratory runs, lobsters have been recorded in a weakened condition. Perhaps when there is good recruitment on the coast, the densities of lobsters preclude many of the migrating white lobsters from finding suitable deep-water habitat, compelling them to keep migrating northwards along the shelf, and for some of them to join the Big Bank Run. Lobsters that survive the Big Bank migration move inshore to shallower waters and join the breeding stock in the northern areas of the fishery.

Once lobsters have been through the migratory whites phase, they become more sedentary, although their nightly foraging sojourns may take them more than a kilometre. Western Rock Lobsters inhabit the same reef, though not necessarily the same den, for long periods. At this stage it remains a mystery exactly how this primitive animal is able to locate its home reef at the end of each night's foraging trip, but underwater observations and radio tracking of tagged Western Rock Lobsters show that they usually approach their resident reef by the same route, suggesting that they develop a good knowledge of the local underwater topography close to their resident surroundings.

Most of the broodstock in the coastal areas is found in depths of 35–90 m; however at the Houtman Abrolhos Islands spawners occur over a wider depth range and are just as numerous on the coral and limestone fringing reefs as in the deep water. These animals show other very different traits to their coastal counterparts. Both sexes mature at a smaller size at the islands and females become sexually mature at about 66 mm carapace length (CL), compared to 90–100 mm CL on the coast. Large numbers of breeding females have remained protected from commercial fishing at the Houtman Abrolhos Islands by their small size at maturity and by the minimum legal size because it is now believed that the broodstock at the Houtman Abrolhos Islands accounts for about 50 per cent of all the eggs produced by Western Rock Lobsters each year. Without that protection to egg production, the heavily fished stock might have produced insufficient larvae to seed the grounds adequately with young lobsters.

REPRODUCTION AND EARLY LIFE

Western Rock Lobsters reach sexual maturity when they are 6–7 years old. Large males are often found with a 'harem' of females under a ledge. The sexually mature population is dominated by females because of management regulations which were introduced in the early 1990s to protect broodstock. Egg-bearing, or 'berried', lobsters have been protected by legislation since 1897. Under a revised management policy introduced in 1992, commercial and recreational fishers must return mature female lobsters (these have long hair-like filaments on their swimmerets), and those above a maximum legal size limit. Initially, there was concern that under this legislation there might be insufficient males to fertilise all the females, but the high percentage of egg-bearing females in the breeding season suggests these fears are unfounded.

Mating takes place during the winter or early spring with the male depositing a sperm packet, or 'tar spot', on the underside of the female behind the last pair of legs. The eggs are laid in the spring and early summer, with small females producing around 100 000 eggs and very large ones up to a million eggs/brood. Most larger females produce two broods each season. When the eggs are being laid on the female's tail, the last pair of legs scrape the tar spot, releasing the sperm which fertilises the eggs as they pass back over the tar spot and the eggs are cemented to the fine hairs on the swimmerets on the underside of each tail segment. The eggs are carried on the underside of the 'berried' female's abdomen for 3–9 weeks (depending on temperature) until the larvae hatch. Hatching, which occurs at night, begins in late spring and continues through the summer.

The newly hatched phyllosoma larvae, rise to the surface of the water, and are carried offshore by surface wind drifts and currents. Offshore transport is rapid, with some larvae moving offshore at speeds of about 5 km/day. The early larvae are weak swimmers in a horizontal direction but they are strong vertical swimmers and make daily movements from the surface to about 60 m depth at dawn and return again at dusk. The main concentration of larvae occurs between 375 and 1030 km offshore, although significant numbers have been found as far as 1500 km offshore in the Indian Ocean.

The larvae remain in the phyllosoma stages for 9–11 months and grow from about 2 mm long at hatching to about 35 mm long in the final larval stage. The range of daily vertical movements in the water increases as the larvae develop and gradually increase their sensitivity to light. The late-stage larvae rise towards the surface at night but usually remain several metres from the surface and can be found at depths of between 60–140 m during the day. The ocean currents off southern Western Australia at these deeper depths flow towards the east and the larvae are transported gradually back towards the coast of Western Australia.

Near the edge of the Western Australian continental shelf, the final-stage phyllosoma larvae undergo what has been described as one of the most dramatic transformations in nature. They metamorphose into the puerulus stage, a post-larval stage intermediate between the larval and juvenile stage. This is very different to the phyllosoma larva: it looks like a transparent miniature lobster and can swim quite quickly (up to 46 cm/sec) and for long distances. This free-swimming stage does not feed but relies only on stored energy reserves and swims 40–60 km across the width of the continental shelf, before settling in holes and crevices in the shallow coastal reefs. It takes 5–11 days to make the trip. A remarkable feat for an animal with a body length of only about 25 mm!

After settlement pueruli darken in colour to blend in with their new surroundings. They are now about a year old, small and very vulnerable to predation. Within about a week of settlement the pueruli moult into the juvenile stage and finally resemble adult lobsters. Juveniles tend to live alone or in small groups for about the first 18 months on the coastal reefs but then join other lobsters and shelter together in holes and crevices where they first are observed by divers.

▽ 'White' and 'red' forms of Western Rock Lobster. *Cathy Anderson*

▷ Ventral view of a female Western Rock Lobster showing the 'tar spot' and eggs. Note also the specialised v-shaped appendage on the base of the fifth leg which is used to 'comb' the eggs, in the process freeing dead eggs from the egg cluster. *Rhys Brown*

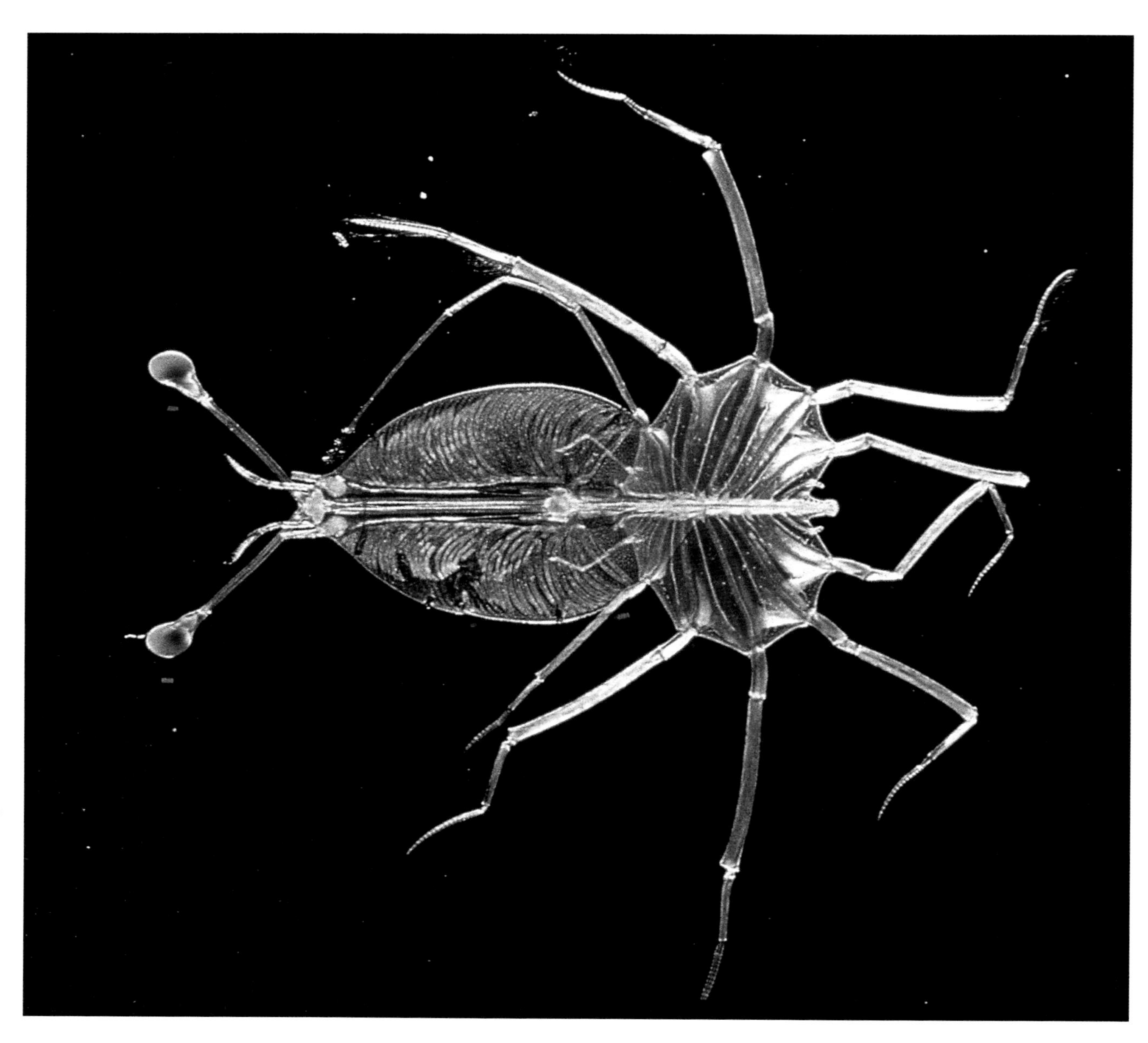

△
A late-stage phyllosoma of the Western Rock Lobster.
Bruce Phillips

The numbers seen by divers are small compared to the total numbers present in the reef systems. In an experimental area at Seven Mile Beach near Dongara, scientists have estimated densities of lobsters around 2–3 years of age of up to 4/m^2. The growth and survival of these lobsters is influenced by the availability of food and shelter, both of which are limited. Many lobsters die in this phase before they can move offshore to the breeding grounds.

Numerous fish and shark species eat Western Rock Lobsters. On inshore reefs the Sand Bass (*Psammoperca waigiensis*) is probably the most common of the fish predators, along with Sea Trumpeter (*Pelsartia humeralis*), Brown-spotted Wrasse (*Pseudolabrus parilus*), Gold-spotted Sweetlips (*Plectorhinchus flavomaculatus*) and Breaksea Cod (*Epinephelides armatus*). These six species may take at least 2500 juvenile Western Rock Lobsters/ha/year. In deeper water, Whiskery Sharks (*Furgaleus macki*), Baldchin Grouper (*Choerodon rubescens*), snappers (Family Sparidae), Chinaman Cod (*Epinephelus rivulatus*) and West Australian Dhufish (*Glaucosoma hebraicum*) are important predators. Octopuses, which occur at all depths, account for a substantial number of lobster losses and often enter commercial pots, infuriating fishers by eating the largest lobsters in the pot. Lobsters are cannibalistic when held under aquarium conditions and may well be important predators of their own kind, particularly in areas where densities are high and dens are limited. However, humans are by far the most important predator of large lobsters. The 600 commercial fishing vessels land an average 10 500 t each season, rising to 13 000 t in good years, while 25 000 recreational fishers account for a further 300–800 t/season. These tonnages represent an annual harvest of well over 20 million lobsters each year.

A monthly index is recorded of the abundance of pueruli returning to coastal reefs from the oceanic environment using collectors composed of 'artificial seaweed' that attract small numbers of settling pueruli as they return to the coast. The numbers of pueruli collected each year reliably predict the catch in the fishery up to four years in advance. The ability to predict future catches is a critical element in managing this important fishery.

There are strong environmental links between the annual numbers of pueruli settling and the ocean climate. The strength of the south-flowing Leeuwin Current affects puerulus settlement levels and, in turn, El Niño Southern Oscillation (ENSO) events in the Pacific (see Chapter 1) influence the strength of the Leeuwin Current. A stronger current and the westerly winds off southwest Western Australia, especially during the peak settlement period in spring, results in higher puerulus settlement.

The correlation between westerly wind strength and puerulus settlement is reversed for recruitment to the offshore Houtman Abrolhos Islands. These islands, about 60 km offshore, are much closer to areas of high larval abundance so the number of pueruli settling is less dependent on their being driven towards the coast by wind and sea conditions. This correlation between the oceanic environment and puerulus settlement clearly indicates that the ecosystem supporting the Western Rock Lobster operates over a very large scale, and that lobsters on coastal reefs are only part of the total picture.

Eddies and tongues of water that develop from the Leeuwin Current as it flows down the Western Australian coast may influence localised puerulus settlement. At the southern extremity of the Western Rock Lobster commercial fishery near Margaret River, settlement occurs on collectors only when the Leeuwin Current is strong. These settlements result in single strong year classes in the lobster population in that region.

Even in those parts of the coast that receive regular puerulus settlement there are large fluctuations among years. Although large settlements translate into large landings once the pueruli become legal-sized lobsters, the differences in catch are not proportional to the differences in puerulus settlement indices. At the Houtman Abrolhos Islands for example, a 50 per cent drop in puerulus settlement results in only an 8 per cent drop in the predicted catch. This indirect relationship between settlement and subsequent catch may be due to density-dependent effects in the population. When lobsters become very dense, food and dens may become limited, which in turn leads to lower survival and slower growth within the population. These effects are more marked at the Houtman Abrolhos Islands than on the coast because of the higher abundance of lobsters there.

FUTURE DIRECTIONS

All over the world, lobster fisheries are being intensively exploited and as a consequence there is increasing international interest in rock lobster aquaculture and grow-out of pueruli to supply the demand. It will be difficult to grow Western Rock Lobsters from eggs to adults because of the complex and long-lived phyllosoma stage. Harvesting pueruli in the wild and growing them to a marketable size under controlled conditions, free from the high mortality rates in the wild, has good potential. Researchers are developing techniques for capturing commercial quantities of pueruli and studying the causes of puerulus mortality. The latter research area is important because it is necessary to predict the impact of large-scale harvesting of pueruli on the wild stocks and the ecosystem.

Research into the biology, ecology and fishery of Western Rock Lobsters commenced in the mid-1940s and has come a long way over the last 50 years. There is still much to learn about Western Rock Lobsters and some of the latest research technology available, such as data-recording tags and underwater video systems, will assist in that process. Aquaculture is an exciting future prospect for this species, and research to develop the means to cultivate lobsters will unquestionably provide a better understanding of this extraordinary animal.

Toxopneustes piledus.
Rudie Kuiter

15 Sea urchins

Neil Andrew and Andrew Constable

Sea urchins are echinoderms, Phylum Echinodermata: they are related to seastars, brittle stars, basket stars and sea cucumbers. Although it may not always be obvious to the casual observer, these diverse animals are united in possessing pentamerous symmetry. This form of symmetry in their body plan is most obvious in seastars and sea urchins in the possession of five arms, five teeth, five roe elements and so on. Even in those seastars with more than five arms there is an underlying pentamerous symmetry.

Sea urchins are structurally simple animals: they are encased in a rigid shell or 'test' of calcium carbonate; they have a simple nervous system but no brain and have a poorly developed musculature. An opened sea urchin is dominated by three structures: the gut, gonad and the complex jaw apparatus known as the Aristotle's lantern. This simplicity of design in the sea urchins offers considerable flexibility in how and where they live and is a key to understanding the enormous success of the group.

DIVERSITY

There are about 20 species of sea urchin in southern Australia, 18 of which are unique to the region. Little is known about the ecology of almost all these species. Research has focused on several species that have a great influence on the ecology of subtidal rocky reefs and in recent times have been harvested commercially. In this chapter we will focus on two species: the Black Sea Urchin (*Centrostephanus rodgersii*) and the Purple Sea Urchin (*Heliocidaris erythrogramma*).

Black Sea Urchins are largely restricted to New South Wales, although they are found west of Cape Conran in eastern Victoria, around islands in eastern Bass Strait and along the east coast of Tasmania. At its northern limit, this species co-occurs with hard corals at the Solitary Islands in northern New South Wales. At their southern limit along the Tasman Peninsula in Tasmania, Black Sea Urchins are found with a typical cold-water

△
Top: Red Sea Urchin.
Ken Hoppen

◁
Left: Purple Sea Urchin.
Rudie Kuiter

△
Right: *Tripneustes gratilla.*
Rudie Kuiter

flora, including Giant Kelp (*Macrocystis pyrifera*) and Bull Kelp (*Durvillaea potatorum*). Populations of Black Sea Urchins have increased in Tasmania in the last 20 years, and were not recorded before 1978. The reasons for this range extension are poorly understood but may relate to changes in the strength of the East Australian Current and associated warming of waters along this coast (see Chapter 4). Black Sea Urchins are most abundant on subtidal rocky reefs not subjected to influxes of fresh water. Floods have killed sea urchins on shallow reefs inside Botany Bay and near the mouths of rivers near Batemans Bay and Eden in New South Wales after heavy rainfalls over short periods.

In New South Wales, Black Sea Urchins are most abundant between Port Stephens in central New South Wales and Wonboyn, near the Victorian border. As with many species of sea urchins, they may be locally very abundant and found in dense aggregations. On reefs in southern New South Wales, densities of more $60/m^2$ have been recorded. At these very high densities they graze patches of the reef back to naked rock. Black Sea Urchins are most abundant in intermediate depths, between 2 m and 20 m. On the margins of reefs in the shallowest water, they occur in crevices and depressions and do not move as far as those in deeper water. At depths greater than approximately 20 m, Black Sea Urchins become less dense and are found in depressions and crevices amongst the sponges and other sessile animals that increasingly dominate reefs at those depths.

Purple Sea Urchins are more widely distributed around southern Australia: from Caloundra in southern Queensland to Shark Bay, Western Australia. Purple Sea Urchins are found in intertidal rock pools and on subtidal rocky reefs. There, they can be found in dense patches in crevices or burrows but on average occur in densities less than $5/m^2$. Purple Sea Urchins can also be found in dense aggregations ($20–80/m^2$) in protected waters, either

△
Diadema palmeri.
Rudie Kuiter

on coarse sandy substrates, within seagrass beds or in deeper waters on reefs. They have a greater tolerance to fresh water than Black Sea Urchins. The name Purple Sea Urchin may be a misnomer in some places because the species may be a wide variety of colours, from violet to green and more rarely white. The proportion of different coloured animals varies among populations and places. The causes of colour variations are poorly understood but are related to the relative concentrations of pigments in the dermis or skin that overlays the test, and in the spines and test. Colouration appears to be controlled by a mixture of genetics and environmental factors, particularly those correlated with wave exposure, although the mechanism is unclear.

FOOD AND FORAGING

The Aristotle's lantern is one of the more remarkable anatomical structures found in the animal kingdom. It is a complex set of 40 skeletal elements and associated muscles that control the movement of the five teeth at the 'heart' of every sea urchin. Sea urchin teeth are capable of biting through the toughest of algae and gouging the encrusting algae and sessile animals that cover much of the reef. These teeth are made of strong calcite and are continually regenerated to replace the ends abraded by rock surfaces. The rest of the lantern helps move the teeth and forms the food into pellets packaged in a mucous covering before it passes to the gut for digestion. This is why, when a sea urchin is cracked open, the gut is filled by balls of food rather than semi-digested fragments of algae.

Sea urchins will eat almost anything but are mostly herbivores and it is by feeding on large brown algae that they have their greatest impact on rocky reefs. Black Sea Urchins consume almost all algae and the patches of Barrens Habitat around their shelters are testament to their efficiency in removing juvenile kelps before they become larger plants. There appears to be some refuge in size for large

△ Black Sea Urchin. *Ken Hoppen*

because Black Sea Urchins do not appear capable of climbing onto large plants. They are, however, adept at trapping and consuming detached plants that drift into the crevices and gutters where the urchins are usually found.

Sea urchins move by the coordinated movement of their spines and a unique system of long extendable suckers or 'tube feet'. If you hold a live sea urchin you will quickly see and feel the many hundreds of tube feet extend and suck onto your hand. This mode of locomotion may seem primitive but sea urchins can move quickly and over long distances during their nocturnal foraging. How they forage and what they feed on is determined by food availability and the type of shelter present on the reef. The availability and type of shelter is determined by the density of sea urchins and rock type.

Around Sydney, where the Hawkesbury sandstone weathers into large undercut benches, Black Sea Urchins typically leave their day-time shelters soon after dark and forage out for distances of up to 10 m from their crevices in search of large brown algae. They return before the twilight of dawn to resume a place in the crevice from which they came. Although they usually return to the same crevice, in what seems to be a form of 'musical chairs' they are rarely found in the same position within the crevice. These limited excursions away from large crevices can cause sharp boundaries between the patches of reef that sea urchins graze and other habitats that contain an abundance of large brown algae.

On reefs south of Wollongong, Black Sea Urchins may be seen out in the open during the day, usually on reefs with high densities of sea urchins and with relatively little shelter. Under these circumstances they are often observed in large aggregations with their long spines interwoven to provide some deterrence to predators such as Port Jackson Sharks (*Heterodontus portusjacksoni*) and Eastern Blue Groper (*Achoerodus viridis*). Their nocturnal foraging behaviour on these more southern reefs has not been well documented but it appears that they do not move great distances, and possibly forage similar distances to those in Sydney. Whether they come home to the same place on the reef is unknown.

Purple Sea Urchins are usually more sedentary than Black Sea Urchins but they can aggressively graze over kelp forests and seagrass beds under some circumstances. They move mostly at night but appear to show little fidelity to single crevices. In intertidal and shallow subtidal waters, these sea urchins remain predominantly in crevices and hollows in the rocks which are formed from many years of scraping with their teeth or from abrasion of their spines. If not in crevices, Purple Sea Urchins can become dislodged by waves and washed ashore on nearby beaches. In these areas, Purple Sea Urchins mostly obtain their food by capturing detached plants drifting past. In estuaries, these sea urchins will graze on anything available, as evidenced by the large quantities of shell grit and sand in the guts of sea urchins living in seagrass beds or on shelly bottoms. When exposed in this way, Purple Sea Urchins are vulnerable to predation by fish, octopuses and large predatory starfish, such as the Eleven-Armed Seastar *Coscinasterias muricata.*

ECOLOGICAL IMPACTS

Although not as colourful or attractive as many animals found on rocky reefs, sea urchins can play an extremely important role in the ecology of reefs. The rise and fall of sea urchin populations has enormous consequences for the types and numbers of algae, fish, and other organisms found on reefs, including those harvested by man. A good example of this effect is the strong negative interaction between Black Sea Urchins and Blacklip Abalone (*Haliotis rubra*) in New South Wales (see Chapter 8). Potentially, the range extension of this sea urchin into Tasmania could have a similar effect on abalone fisheries there, and potentially even fisheries for Southern Rock Lobsters and reef fishes.

In most of the temperate seas of the world sea urchins can remove large brown algae from areas of reef and thereby fundamentally change the ecology of the reef. The ecological role of sea urchins and their impact on kelp forests over large areas of reef has been the focus of research on the east and west coasts of North America, New Zealand, South Africa, Chile, Japan, and New South Wales. Not all species of sea urchin have this dramatic effect on the abundance of large brown algae, and some species change their behaviour on a set of circumstances that are poorly understood. Sea urchins in the northern hemisphere, genus *Strongylocentrotus,* can swap from a sedentary existence feeding on drift algae to a mobile habit in which they form large aggregations that move through kelp forests eating all the large brown algae.

In New South Wales, grazing Black Sea Urchins create and maintain the Barrens Habitat in which all large brown algae have been removed, leaving only encrusting algae, limpets and sessile animals, particularly encrusting sponges. The Barrens Habitat covers about 50 per cent of nearshore reefs in central and southern New South Wales and thousands of hectares of reef. Within this habitat the density of Black Sea Urchins may vary between 5 and 60/m^2. Their ecological impact, in terms of the maintenance of patches of Barrens Habitat is broadly similar across this density range because small kelp plants are vulnerable to grazing and are removed before they establish themselves. Across this density range there are subtle differences in the density of limpets and smaller algae. Areas of Barrens Habitat vary between the small patches surrounding single sea urchins to many hectares that leave only a thin band in the shallowest areas of reef. In contrast to other species of sea urchin, such as the Purple Sea Urchin described below, Black Sea Urchins do not appear to form aggregations that remove all large algae in their path. A consequence of these limitations in behaviour is that the locations and boundaries of patches of Barrens Habitat are more stable through time.

Unfortunately many divers do not understand the important ecological role played by Black Sea Urchins and they feed them to fish, particularly Eastern Blue Groper. The end result of many people doing this can be a fundamental change in the ecology of reefs. 'Feeding the fish' underwater may be fun but it degrades the environment divers come to enjoy.

No other species of sea urchin in Australia plays such a dominant ecological role in the ecology of rocky reefs as the Black Sea Urchin in New South Wales. In all the other states, sea urchins play only a minor role in the ecology of exposed rocky reefs, although they may have occasional and unpredictable impacts within estuaries. Although present, they are generally in low densities and do not appear to actively forage over the reef, instead making their living by feeding on drift algae. Purple Sea Urchins can have similarly dramatic effects on large brown algae and seagrass beds in sheltered waters. Very large densities (up to 180/m^2) of Purple Sea Urchins can form 'fronts' on the edge of seagrass beds in some protected waters. In Corio Bay, Victoria, some fronts appeared relatively stationary in the 1980s and were observed to remain so for many years. In contrast, episodic 'outbreaks' of Purple Sea Urchins in the seagrass beds in Botany Bay have prompted ill-fated attempts to control the species.

GROWTH AND AGE

The broken tests of sea urchins are a common sight on beaches near rocky reefs in southern Australia. Inspection of these tests shows they are made up of many small plates sutured together. Sea urchins are able to respond to poor food availability by resorbing the calcium carbonate in the test and actually shrinking. The ability to do this differs among species and provides considerable flexibility in how they respond to limited food. Many of the habitats sea urchins live in are, by their own making, largely devoid of red and brown algae, their preferred food. Sea urchins are able to tolerate these food-limited conditions because they lack large muscles and have a simplicity of design that allows them to slow or stop their growth and reallocate energy from reproduction to maintenance. These features of their biology allow sea urchin populations to persist on reefs in the absence of their preferred food for many years.

Given the great plasticity in growth, it is unsurprising that size can be a poor indicator of a sea urchin's age. Fortunately, growth can be studied in sea urchins by marking the bones of the Aristotle's lantern with the antibiotic oxytetracycline. A small amount of oxytetracycline injected into sea urchins at high concentrations leaves a fluorescent mark, as it did on the teeth of babies given antibiotics in the 1960s. As the sea urchin grows, this mark effectively date-stamps the size of the urchin at tagging. Estimates of annual growth can be derived by measuring the distance between the mark and the growing edge of the main bones of the lantern (the demipyramids) one year later.

Annual growth checks appear naturally on the demipyramids of Black Sea Urchins and can also be used to estimate growth. The trick is to ensure these marks are annual or at least follow a predictable cycle. In New South Wales, recent validation studies appear to confirm that the marks are annual and they are now being used to age this species, at least in relatively young sea urchins. Being able to age sea urchins in this way allows a much wider range of research tools to be used in describing their demography and assessing the fishery. This advantage will be important if the fishery for Black Sea Urchins develops further in New South Wales.

Black Sea Urchins grow relatively quickly, reaching approximately 40 mm test diameter in their first year of life. Growth thereafter slows and by the time a Black Sea Urchin is 80 mm across the test it may be between five and ten years old. Individuals larger than 100 mm test diameter may be as old as 15 years. Purple Sea Urchins are generally smaller than Black Sea Urchins but, in Tasmania, they can grow to 125 mm test diameter, although the most common size encountered throughout their

▽ Black Sea Urchins at high densities in a Barrens Habitat. Eden, New South Wales. *Nokome Bentley*

range will be 60–90 mm in diameter. These more common animals grow more slowly than Black Sea Urchins and are about 6–8 years in age at full size.

The growth and body shape of Purple Sea Urchins depends to some extent on where they live. In habitats with poor food supply, such as in patches of Barrens Habitat, they have proportionally larger Aristotle's lanterns, less spine mass and, in younger urchins, more spherical tests. The size they reach can be limited in areas exposed to wave action. Here, sea urchins are easily damaged and as a result, the test and spines of Purple Sea Urchins are thicker compared to those in more protected habitats. These changes in morphology mean that growth in overall size slows because calcite is being diverted to strengthening the test and spines rather than expanding test size.

REPRODUCTION

Compared to many of the organisms described in this book, sea urchins have relatively simple reproduction. Aside from a few rare individuals, sea urchins do not change sex. Males and females broadcast their eggs and sperm directly into the sea. Fertilisation success depends on adequate densities of sea urchins being in close proximity and all spawning at the same time. The processes controlling the timing of spawning are poorly understood. Although brooding of young has been recorded for some species, particularly those in deep water and in high latitudes, this is relatively uncommon. In those species that do brood, juveniles are retained under a protective spine canopy on the membrane around their mouths or in pouches on the outside of the female's test.

Reproduction in sea urchins is complicated by the fact that the gonads (the roe) are the principal nutrient storage organs. The size of the roe may therefore increase in response to both the annual reproductive cycle and to increased food availability. Under poor food conditions, output of eggs or sperm can be small and in some instances reproduction may not occur at all. In the Black Sea Urchin, low availability of food in the Barrens Habitat means that the roe of urchins in this habitat is usually less than half the size of those in habitats with an abundant supply of food. Further, poor food quality and quantity means that the annual cycle in roe size is much less apparent in the Barrens Habitat than it is in the Fringe Habitat. In the Purple Sea Urchin, the same differences are observed. In both species, the roe can rapidly grow if a large influx of food occurs, either as drift algae or if algae settle and grow near the sea urchins.

Each year, Black Sea Urchins begin to spawn in June, at a time of year with short days and lunar conditions coinciding with the winter solstice. At the Solitary Islands, they have a short one-month spawning period but in the southern parts of their range breeding lasts five to six months. These differences in the duration of the spawning season are probably caused by water temperature. At more southern locations,

▽ Purple Sea Urchins at high densities can remove large seaweeds and form areas of Barrens Habitat in the same way as do Black Sea Urchins. Portsea, Victoria. *Rudie Kuiter*

the development of eggs and sperm is not as synchronised as in the north and the reproductive season is marked by a series of partial spawnings. The cues for spawning within this long season are not well understood but are thought to be related to lunar cycles and sea conditions. Once fertilised, eggs and larvae drift in the sea for approximately a month. During this time they develop through several larval stages that differ enormously in appearance. The currents and tides will largely determine the numbers of larvae arriving on a reef to settle but beyond this general consideration, the processes that determine when and where Black Sea Urchins settle back onto reefs and survive to adulthood are poorly understood.

Purple Sea Urchins differ from the Black Sea Urchins reproductive patterns described for Black Sea Urchins in several important ways. Firstly, and unusually for sea urchins, they have heavily yolked eggs. As a result, the larvae do not need to feed in the plankton and fully develop in only a few days. Larvae may then settle back on the reef in habitats near to their parents. Second, the gametes mature later in the year and are ready for spawning around November. Whether spawning occurs in a single event or on numerous occasions over summer is not clear but evidence suggests that spawning behaviour may be similar to Black Sea Urchins. Mature roe ready to spawn can be found in Purple Sea Urchins between November and March. Recent spawning can be observed on the surface of the water because the large eggs float and can be seen to form reddish orange slicks if the water is calm.

△ Purple Sea Urchins are often found in a variety of colours including white with purple-tipped spines. Western Australia. *Rudie Kuiter*

△
Black Sea Urchins feeding on drift Bull Kelp. Green Cape, New South Wales. *Ken Hoppen*

FISHERIES

The roe of sea urchins is a valuable seafood in many parts of the world, particularly Japan and France. Japan dominates the world market for sea urchin roe where the neat wooden trays of 'uni' can fetch more than A$150/kg. Worldwide, about 60 000 t/year of sea urchins are harvested but this is not sustainable and for this reason, aquaculture of sea urchins has been a hot topic for research and development. Most of this research has focused on improving the yield and quality of roe and has included the development of artificial diets. Roe quality in wild sea urchins is determined by a mixture of diet (and related factors such as habitat and density) and reproductive condition. At their peak, roe make up more than 15 per cent of the weight of the sea urchin. Roe quality is poorest during and immediately after spawning.

Sea urchin fisheries are in their infancy in Australia but are likely to expand as fisheries in other parts of the world collapse. A fishery for the Purple Sea Urchin has existed in Tasmania since 1980, primarily on the east coast. Most of the commercial catch has been taken in relatively sheltered waters on rocky reefs (up to 5 m deep), although the catch now mostly comes from shelly bottoms in waters deeper than 10 m. The roe is sold fresh to local traders at around A$42/kg and subsequently auctioned in Japan if harvested in the prime season in the later months of the year.

A fledgling fishery for Black Sea Urchins exists in New South Wales but variable roe quality and poor yields have slowed its development. Perversely, most Black Sea Urchins live in the Barrens Habitat and their small and uniformly low-quality roe attract a poor price. Current research focuses on enhancing roe quality through a better understanding of the relationship between sea urchin density and food availability. Given the demonstrated importance of this species to the ecology of subtidal reefs, and their impact on the abundance of Blacklip Abalone (see Chapter 8), great care will be required in developing management plans for the sea urchin fishery in New South Wales.

Sessile animals crowd the reef in deeper water. Rottnest Island, Western Australia. *Peter Morrison*

16
Sessile animals

Michael Keough

DIVERSITY

When you move below the canopy of a kelp forest and look at more shaded areas, or into deeper water on the margins of reefs, you encounter animals very different from those introduced elsewhere in this book. First, many of them do not even look like real animals at all; they are variously sheet-like and folded into strange shapes or massive tree-like structures, and often gaudily coloured. The subjects of all of the photos in this chapter (and in the Rocky Reefs Beyond Most Divers Box in Chapter 2) are sessile animals — that is, they have an adult form that is attached permanently to the reef and a mobile larval stage. On many reefs along Australia's temperate coastline, these animals cover most of the space available, particularly in more sheltered and deeper waters. They grow over rocks, shells, and pier pilings, and even attach to each other. The underlying reef is often not visible, being completely covered by a colourful mosaic of differently shaped sessile animals. Although they look strange, they are an important part of what makes Australia's temperate rocky reefs special.

Sessile animals of rocky reefs come from many different taxonomic groups. Principally they are sponges, ascidians, bryozoans, hydroids and, to a lesser extent, corals (see pages 140–41 for an introduction to these groups). Sessile animals from southern Australia are more diverse than in other temperate regions of the world. The region is made more special though because many species living here are endemic, or unique. As an example, there are more than 500 species of sponge in southern Australia, 60 per cent of which are endemic and even within the region many of these have restricted distributions. The same is true of more than two-thirds of the species of ascidians (around 200), and many of the 500 or more species of bryozoans found in southern Australia.

△
Bryozoan, *Celleporaria* sp. and red gorgonian coral *Mopsella zimmeri* at Stenhouse Bay, South Australia. *Michael Keough*

Sessile animals are perhaps the most poorly understood group of marine animals. Historically they have had little commercial value and they do not enjoy the familiarity we have with many species found on rocky reefs. The catalogue of species and appreciation for their ecology will surely grow as further studies are done. Recent discoveries of pharmaceutical compounds in many sponges and ascidians will provide a boost to this process. Sessile animals in many ways exemplify the diversity and uniqueness of the marine animals and plants of southern Australia.

MODULARITY, FLEXIBILITY AND LONG LIFE

The world of sessile animals is very different to that of the more familiar marine animals such as fish or crustaceans. In order to understand how sessile organisms live, several concepts have to be introduced. The key feature of most of these animals is that they are modular — what appears to be a single organism is in fact a colony of many smaller units, or modules. These modules are called zooids (in ascidians and bryozoans), or polyps (in corals and hydroids). In species such as soft and hard corals and the colonial ascidians, the individual modules can be seen easily underwater. In others, such as bryozoans, zooids are tiny (usually 0.5 mm long), and can only be seen clearly under a microscope. These organisms grow and reproduce using a mixture of sexual and asexual reproduction. Zooids and polyps are produced asexually by cloning perfect copies of themselves. In other phases of their life cycle they reproduce sexually (by fertilising eggs and producing a planktonic larva). The larva settles, attaching itself to a hard surface, and then begins reproducing asexually, adding new modules which are all genetic copies of the original larva that settled — they are clones.

Individual modules are often capable of living independently, as they do in some anemones, such as the common species, *Anthothoe albocincta.* Most of the time, though, the modules are connected to, and exchange nutrients with other modules, and form a discrete colony. As these colonies grow they can assume a wide variety of shapes, from long single chains to large 'coral-like' organisms, such as gorgonian corals. The modules can join together in two-dimensional sheets, such as in many bryozoans and ascidians.

Complex shapes can arise if, for example, modules growing in sheets fold back on themselves, producing strange scalloped colonies.

This modular construction also allows colonies to recover from damage — if a storm or a predator removes half the colony, it can rebuild itself, because the colony is really just the sum of its parts, the individual modules. The colony could even break into genetically identical fragments, with each fragment living and growing independently. These complexities in biology mean that the genetically unique individual represented by the original larva may be very long lived. Although the original module might have died, genetically identical copies may survive and keep reproducing. Some of the oldest living organisms are in fact modular, partially because of the great flexibility afforded by their capacity to reproduce asexually in response to damage from storms, predation, competition and other hazards. A fish or an abalone that has been cut in half by a predator would not fare so well.

△ The common sea anemone *Anthothoe albocincta*. Southeastern Tasmania. *Simon Talbot*

As a result of breakage and regrowth, the size of a colony does not provide any reliable clues to its age. A small colony could be either very young, or a piece of a larger colony that has been overgrown or broken. Similarly, a large colony could have grown steadily since the larva settled and metamorphosed, or it could have shrunk and fragmented and re-grown a number of times. Being broken or fragmented frequently often comes with a cost, however, because fragments generally do not grow and reproduce as well as the intact colony. In many species, small 'colonies' tend to grow fast but larger colonies reproduce much more. In other species, fragments grow slowly and they do not reproduce as much. This enormous potential for adapting to local conditions and the diversity of patterns in grow-ing, reproducing and dying, make modular species much more difficult to study than 'simple' organisms like fish, which move through a predictable series of life stages, and (rarely) fall to pieces! Many sponges are a little different in having more organisation in the 'colony'. Although these animals do not have 'organs' as higher animals do, they are best considered a single animal.

COMPETITION

Competition for space is one of the most important ecological interactions among these animals. They all need space to attach and grow — a bryozoan or ascidian that covers a large area will have many zooids, each of which is capable of producing larvae, and contributing to the next generation. In many species, larger colonies also have a lower mortality rate than small ones. Despite the clear advantages of being larger, there is often very little visible open space to grow into, so the only way to get larger is to grow over your neighbours. Modular animals are particularly good at growing over other organisms — their flexibility of growth allows them to grow around their neighbours, encircle them, and eventually cover them. The 'victim' can also react, growing away from the site of competition by producing new modules on sides where there is free space, or stopping growth at the edge where the competitor is. Sometimes, there might be retaliation where a bryozoan colony is getting smothered by an ascidian, but has managed to grow up over the ascidian. The result of these interactions is often a rock surface that is a mosaic of colonies of all shapes and sizes. This mosaic changes constantly as competition goes on, but there seems to be no consistent winner.

This competition for space has been studied in detail at places like Edithburgh, in South Australia, and Portsea, at the entrance of Port Phillip Bay, in Victoria. In general, ascidians and sponges are better competitors for space than bryozoans, which are better than unitary or solitary animals like barnacles and tube worms. Some ascidians and sponges seem to win and lose almost at random. It may be that slight differences in the rock surface are enough to change the outcome of competition — if species A happens to be growing on a slightly higher section of rock than species B, it may be able to overgrow B, but if B is on a small bump in the rock, B may have the

SESSILE ANIMALS OF ROCKY REEFS

The sessile animals share the features of requiring some hard surface for attaching themselves, and they feed by extracting food, usually as plankton, from the water around them. The major groups of sessile animals on southern shores are as follows.

Sponges *Phylum Porifera* are very primitive multi-celled animals, lacking well-developed nervous systems. They come in a great variety of shapes and colours, and are often rich in chemicals that could deter predators or help them in competing for space. There are many species in southern Australia, but they are very difficult to identify.

Hydroids, anemones, sea pens, sea whips and soft corals *Phylum Cnidaria* are true modular animals that often form upright branching colonies, with the individual modules, or polyps, along the branches. The polyps have simple body plans, with a ring of tentacles surrounding their mouth. The polyps can be microscopic, or as 'big' as a centimetre or so. Reef- building corals in the tropics, and anemones also belong to this group.

Worms *Phylum Annelida.* One of the earliest colonisers of any open space are tube-building worms, especially serpulid worms, which build calcified tubes. Later, there may be feather-duster, or sabellid, worms, which have soft, sandy tubes. These worms are solitary.

Bryozoans *Phylum Bryozoa* are all modular, made up of small zooids, each enclosed in a calcified or chitinous box (see earlier photograph this chapter). Individual zooids may be specialised for feeding, defence and reproduction. They feed with a specialised ring of tentacles, called a lophophore.

Barnacles *Phylum Crustacea* are solitary, shelled crustaceans. They are typically some of the earliest colonisers of bare space, and do not compete well.

Ascidians *Phylum Chordata* are invertebrate animals that are close, in an evolutionary sense, to vertebrates. The adults, which can be solitary (eg, Cunjevoi, *Pyura stolonifera*) or modular, forming large colonies. Solitary species are usually encased in a test or tunic — a fleshy coating that can be many centimetres thick. The adults are simple filter feeders, but they produce a larval stage that looks like a tadpole, with a long tail. The larval stage has a well-developed nervous system, and other features that let us see the relationship of this group to vertebrates. They are very diverse in southern Australia.

◁
A large grey sponge (*Lotrochota* sp.).
Sue Morrison

▷
Fan or feather-duster worms extend their feeding tentacles above their neighbours.
Ken Hoppen

▷
Zooanthid, *Zooanthus robustus* colony.
Michael Keough

◁
Sea Whips.
Matt Edmunds

▷
Colonial ascidian, *Botrylloides magnicoecum*. The individual zooids are seen in parallel rows. Water is filtered in through the zooids and out through the central siphon.
Ken Hoppen

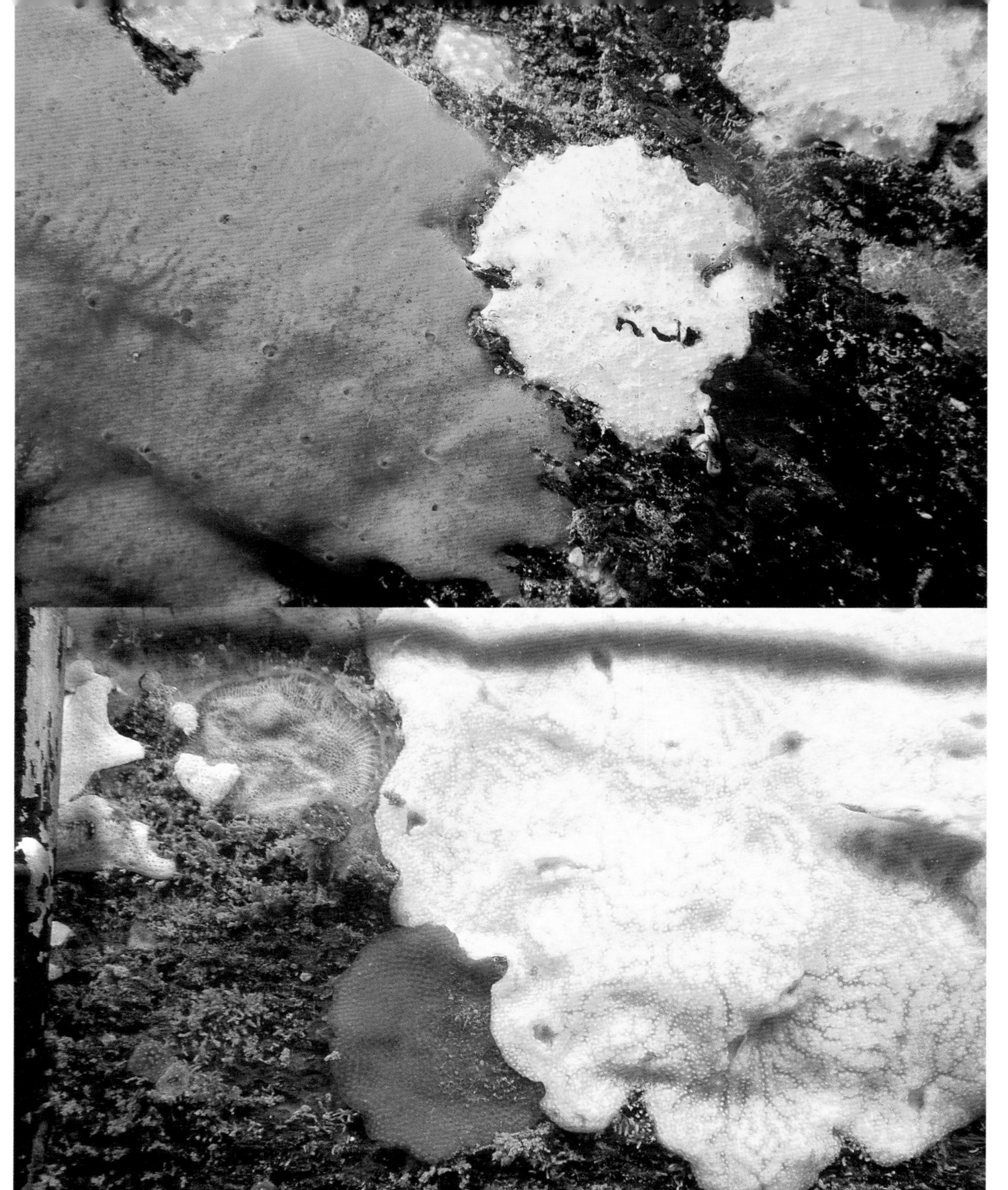

◁ Two sponges competing, with one beginning to encircle the other. Edithburgh, South Australia. *Michael Keough*

◁ The bryozoan *Mucropetraliella ellerii*, being overgrown by a colonial ascidian at Queenscliff, Victoria. *Michael Keough*

advantage. We do not really understand what decides the outcome of these competitive interactions, but we do know that there is not a single species able to outcompete all the others.

SURVIVING WHEN YOU'RE NOT A GOOD COMPETITOR

With space at a premium, life can be difficult for any species that does not compete well — what happens to bryozoans if they loose out to sponges and ascidians, or tiny tube worms that are outcompeted by everybody else? These weak competitors are forced to exist by either taking advantage of areas of bare space as they become available, or by attaching themselves to surfaces that are not used by other species, such as mollusc shells or kelp fronds. Along the southern coast, you may also see small scallops whose shells are almost always covered by sponges and bryozoans and, living out in sandy areas, you may see beds of Razor Shells (*Pinna bicolor*) which are good places for bryozoans, barnacles and tubeworms, because they rarely encounter sponges and ascidians there. Razor Shells are particularly common in the two Gulfs of South Australia.

On rock walls and pier pilings, though, the important events are those that clear space — storms tear organisms away, predatory fish and seastars can create small 'gaps', and sometimes the animals just get old or diseased and die. Whenever this happens, there is some unoccupied space for a short while. These gaps get filled quickly, as some of the modular organisms alongside spread into the space, and planktonic larvae arrive, settle and start growing. A bryozoan or barnacle larva that happens to be in the right place at the right time may settle, grow and reproduce before being overgrown eventually but there is a much smaller chance of these larvae successfully colonising gaps. Despite this, many bryozoans and tube worms have an advantage over the colonial ascidians and sponges because they usually reproduce more often and produce lots more larvae.

An important feature of the larval stage of many of these sessile animals, especially the modular ones, is that larvae spend only a very short time in the plankton. Many familiar animals, such as fish, lobsters and barnacles, spend a comparatively long time in the plankton. The larvae of some fish and rock lobsters may spend more than a year in the plankton. Almost all bryozoans and colonial ascidians, and many other modular sessile animals, such as sponges, have larvae with no feeding structures that spend as little as a few minutes, and usually less than one day, as part of the plankton. Even some of the common solitary animals, such as ascidians have larvae that last only a few days. Having a short-lived larval stage is important because the larvae do not travel great distances unless there are extremely strong currents. For most of the species that we see in these communities, travel between rocky reefs may be very difficult, and not happen very often. Areas that are close together, such as rocky reefs a few kilometres apart, may have quite different sessile communities, perhaps because the wave exposure, depth and so on differs. More importantly, the species found in each place may just be the result of historical 'accidents' of colonisation. Subsequent restricted larval dispersal will keep these areas different from each other. Within a single reef area, we will also find that individual clearings, shells and rocks will be colonised by different groups of species.

Having a short larval period may also be important for the evolution of these organisms. In the past, we tended to think of marine species as having wide ranges, with their genetic material mixed thoroughly by larvae that travelled long distances. When the larvae do not travel far, local populations may become genetically isolated from each other, and, over time, quite different. If the reefs, jetties and marinas are close together, they might act as stepping stones, allowing exchange of larvae over wide areas, but as the spacing between areas gets bigger, dispersal becomes much more difficult. As we bring sophisticated new genetic tools to bear on these sessile animals, we are finding that many of them do seem to consist of small, isolated populations. These are just the conditions which we might also expect to be favourable for the development of new species, and some of the species that we currently recognise may turn out to be whole sets of species with small ranges.

CASE STUDIES IN DEPENDENCE

The limited dispersal of larvae and intense interactions among colonies that are played out over small scales has meant that many unusual relationships among species have evolved. Two such relationships illustrate the special circumstances of some of these animals. The first is between a bright red bryozoan, *Mucropetraliella ellerii* and its predator, a nudibranch named *Madrella sanguinea.* The second is among two ascidians (*Cnemidocarpa pedata* and *Pyura spinifera*) and the sponge that lives on them.

MUCROPETRALIELLA ELLERII AND *MADRELLA SANGUINEA*

The bryozoan *Mucropetraliella ellerii* is common on rocky reefs and pier pilings in southern Australia. Colonies grow in a wide variety of places, including the stems of red algae, and on the tunics of large solitary ascidians. It is also one of the few bryozoans to also occur in the intertidal zone of rocky shores — you can find it under boulders and under ledges near the low-water mark on rocky reefs in central Victoria, across to parts of South Australia. *M. ellerii* probably occurs in a wider range of habitats than any other bryozoan. When we think about the range of places this animal occurs, we can see some of the advantages of being modular. Living on solitary ascidians and seaweeds can be difficult because they are flexible and can shrink and expand. On the other hand, if few other species can live in such places, they may be good places for a relatively poor competitor to live. *M. ellerii* colonies are not firmly cemented to the surface, but attached by a series of tiny rootlets, providing some flexibility.

The tiny nudibranch, *Madrella sanguinea,* is less than 1 cm long, and lives and feeds on colonies of *M. ellerii.* It belongs to a family of nudibranchs that specialise on bryozoans. Nudibranchs and other gastropod molluscs, such as abalone and limpets, feed with a special structure called a radula, which is a chitinous ribbon, bearing a variety of teeth. This radula can be moved out through the mouth, and scraped over surfaces to remove algae and other food. It has become highly specialised in many molluscs, and in *M. san-*

guinea the radula has a giant, curved tooth, which is used to tear open individual zooids of the bryozoan colony. The nudibranchs feed on the colony and, if it is a large or fast-growing colony, the bryozoan may produce new zooids as fast as the nudibranch can eat them, so the nudibranch has a reliable food supply. If the prey colony is too small or too slow growing, it may eventually die, and the nudibranch will need to move to another colony. While it is on the bryozoan, it is very hard to see, being almost the same colour and texture, but the nudibranchs are conspicuous when they move between colonies. They are able to distinguish between different species of bryozoan, and feed only on *M. ellerii.* We do not know whether they can detect a colony of *M. ellerii* from a distance, or whether they need to make contact with the colony to identify it.

Reproduction can also be a problem — with larvae dispersing through the plankton, how do they find their prey successfully? These nudibranchs have solved this problem by simply bypassing the planktonic larval stage completely. The parent nudibranch lays a spiral ribbon of jelly, with large (about 0.3 mm) fertilised eggs. The eggs are dark red, and very hard to see against the red bryozoan. Rather than moving into the plankton, the larvae develop within this jelly ribbon, and metamorphose from larvae into tiny nudibranchs, and then crawl out onto the bryozoan. This is a solution with some risks — the babies will find food, but if they all stay and feed on their 'birthplace', they will run out of food, and must eventually move off and find more colonies. It may be that finding new places to eat is less risky as an adult than as a larva or baby nudibranch.

CNEMIDOCARPA PEDATA AND ITS ASSOCIATES

Although most interactions between neighbouring sessile animals are intensely competitive, this is not always the case. In southeastern Australia, and in New South Wales in particular, clumps of bright purple or yellow stalked ascidians are relatively common in sheltered bays and in deep water on exposed coasts. These ascidians are attached to the rock surface by a small holdfast, with the main part of the animal at the end of a stalk that can be over 30 cm long. These striking animals are in fact part of a clump of three species: the stalked ascidian (*Pyura spinifera*), an unstalked

▷ Stalked ascidians (*Pyura spinifera*) growing from a solitary ascidian (*Cnemidocarpa pedata*). The purple colour of the stalked ascidians is provided by a sponge, *Halisarca laxus*, that overgrows both species. Bass Point, New South Wales. *Rudie Kuiter*

◁
The bryozoan *Mucropetraliella ellerii*, with the predatory nudibranch *Madrella san guinea* crawling over it. Portsea, Victoria. *Rudie Kuiter*

solitary ascidian (*Cnemidocarpa pedata*) and a sponge (*Halisarca laxus*) which gives the clumps their characteristic colour.

The clumps are seeded by the solitary ascidian *Cnemidocarpa pedata* which settles directly on to the reef surface. Larvae of the stalked ascidian *P. spinifera* settle from the plankton on to these ascidians, seemingly cued by chemicals in the water around *C. pedata*. As they grow, the holdfast of the stalked ascidian establishes itself on the reef adjacent to its host ascidian and clumps form as more *P. spinifera* arrive and settle. The clumps may grow to include up to 22 *P. spinifera*. These stalked ascidians have a seemingly obligate relationship as they are only rarely found in the absence of *C. pedata*.

The third species in this relationship is the encrusting sponge *Halisarca laxus* which in turn has an obligate relationship with the ascidians. *H. laxus* has only ever been observed on these two species of ascidian. Clumps of ascidians are covered by a sponge that is a clone from a single planktonic larva, which settled, grew and reproduced asexually. The sponges on the nearby clumps of ascidians will be genetically different from the first clone, telling us that it originated from a different larva. This set of interdependencies reveals some of the complexity and the diversity possible among sessile species. Further, it provides a clue to the potential importance of water-borne chemicals to the ecology of rocky reefs.

CONCLUDING REMARKS

Communities of sessile animals are very dynamic — organisms overgrow each other constantly, clearings get made, and planktonic larvae colonise the clearings. Some of those larvae produce thriving colonies, while others are overgrown quickly. If you visit the same area of rocky reef at different times, you are unlikely to see the same pattern of sessile animals — the same species may still be present, but their sizes, shapes and spatial arrangement may be completely different. The limited dispersal of many of the species will mean that rocky reefs quite close to each other will have very different faunas. These communities are an important part of Australia's marine diversity, and may need to be managed differently from the more familiar mobile marine species.

A female Grey Nurse Shark cruising over the reef. South West Rocks, New South Wales.
Ken Hoppen

PART 3
Sharks, fishes and seals

Egg cases of a Crested Horn Shark. These eggs have been laid among a cluster of ascidians. The long tendrils from the apex of each egg case secure the eggs to the ascidians. Jervis Bay, New South Wales. *Kelvin Aitken*

17 Reef sharks and rays

Marcus Lincoln Smith

In many ways, humans associate sharks with the darker nature of the sea — menacing, restless and unpredictable. Nowhere is this more evident than in TC Roughley's (1955) classic book, *Fish and Fisheries of Australia*, and the illustration by Fraser-Brunner in the 1955 edition showing a Great White Shark (*Carcharodon carcharias*) cruising along the edge of a reef. The shark's pectorals are arched aggressively, the reef is bare and more like a mountainous crag than a true reef, the watery backdrop is pale green and grey like an impending storm and, cradled between two rocks, is a human skull. Roughley's book, first published in 1951, was the pre-eminent book on the fisheries of Australia at the time and reflected the attitudes of the day. Since those days attitudes toward sharks have evolved from fear to exploitation and to a sense of mastery over these animals. While we still feel the need to protect ourselves from sharks through the use of mesh nets across beaches, there has been a strong push in recent times to conserve these animals.

In Australia there are major ecotourism ventures based around viewing Whale Sharks (*Rhincodon typus*) off northern Western Australia, Great White Sharks off South Australia and Grey Nurse Sharks (*Carcharias taurus*) in northern New South Wales. Moreover, both White Sharks and Grey Nurse Sharks (and the closely related Herbst's Nurse Shark, *Odontaspis ferox*), are now protected from exploitation in all or parts of Australia. There are many reasons for this evolution in attitude toward sharks, including the development of scuba diving and an increasing popular awareness of the importance of species other than humans. There has also been a great increase in our knowledge of the biology and behaviour of sharks. With this knowledge comes the discovery that sharks are far more vulnerable to the activities of humans than we are to that of sharks.

◁
Port Jackson Shark resting the reef. Note the distinctive black harness pattern between the dorsal and pectoral fins. Jervis Bay, New South Wales. *Kelvin Aitken*

◁
A Crested Horn Shark swimming over rocky reef. Note the high crest above and behind the eyes and the blotchy colouration along the body. Montague Island, New South Wales. *Ken Hoppen*

It is perhaps ironic that we know a great deal about some aspects of the biology and ecology of sharks, but almost nothing about other aspects. Sharks are often relatively easy to capture, either in nets or by lines, and scientists have had an abundant supply of specimens to provide us with knowledge on reproduction, anatomy, size, feeding and, to some extent, distribution of species. Ready capture has also led to numerous tagging studies and we know that some sharks, such as Blue Sharks (*Prionace glauca*), swim vast distances and that many are quite long-lived (at least several decades). Unfortunately, however, it is difficult to determine the exact age of sharks because their skeletons are made of cartilage, which can be difficult to age. Another good source of information has been behavioural observations of sharks and rays held in aquaria. Some sharks, notably the sluggish bottom-dwelling species, adapt well to captivity and may be kept for many years. Observations of these species have yielded information on feeding rates and diet, reproduction and interactions with other species. These observations are limited, however, because we do not know if sharks respond to stimuli in similar ways in nature.

Despite these sorts of information, there are still major gaps in our understanding of the ecology of sharks, particularly in regard to migration, small-scale movements and the interactions between sharks and the reef environment. Moreover, whilst much of modern aquatic ecology uses manipulative experimentation to determine the processes that structure the flora and fauna of reefs (eg, clearing patches of kelp, or moving sea urchins to measure responses), most sharks are not amenable to this type of research because they are too rare, too large or too mobile. Researchers have therefore tended to rely mostly on observational studies, often with some quite elegant refinements to tagging. This chapter summarises information on three varieties of sharks and one group of stingrays that divers are likely to encounter on shallow reefs in southern Australia. These varieties include the horn sharks (Family Heterodontidae), wobbegongs (Family Orectolobidae), Grey Nurse Sharks, and stingrays and stingarees (Families Dasyatidae and Urolophidae).

HORN SHARKS

Eight species of horn sharks are recognised worldwide. These sharks are all relatively small to moderate sized (about 1–1.7 m long) and all share a similar body form, including the presence of an anal fin plus a sharp spine in front of each dorsal fin. All species have the familiar short, blunt snouts and teeth that vary markedly from the front of the mouth (where they are small and have cusps) to the back (where they look like molars). This difference in tooth form has lead to the family name, Heterodontidae, meaning 'different teeth'. This dentition serves them well; they generally feed at night on a wide variety of reef- and sand-dwelling invertebrates, including sea urchins, marine snails, crabs and occasionally small fish. The small, frontal teeth are used to grip prey, while the molars can crush shells.

In temperate Australian waters there are two species of horn shark: Crested Horn Shark (*Heterodontus galeatus*) and Port Jackson Shark (*Heterodontus portusjacksoni*). The Crested Horn Shark occurs from subtropical Queensland to New South Wales and possibly Tasmania. The Port Jackson Shark extends from southern Queensland, New South Wales and Tasmania across the Great Australian Bight and north along the coast of Western Australia to the Houtman Abrolhos Islands. There is also a single record from New Zealand. Analyses of the genetics of these sharks suggests that there are two separate stocks of Port Jackson Shark, one occurring from Queensland to Jervis Bay (with migrations to Bass Strait), the other extending from Victoria to Western Australia. Within the eastern stock, there may be distinct sub-populations centred on Newcastle, Sydney and Jervis Bay.

The two species look fairly similar and can be confused. Port Jackson Sharks have a triangular, black, harness-like set of bars converging on the sides and back from the origin of the first dorsal fin and bases of the pectoral and pelvic fins, and a black vertical stripe through the eyes. They also have a very low dorsal profile behind the eyes. Crested Horn Sharks are readily identified by the high ridge behind the eyes and the broad, black bands on the head and the back at the base of the first dorsal fin.

Where they lay their eggs and the appearance of the eggs may also be used to distinguish the two species. Crested Horn Sharks lay their eggs in deeper water (eg, >20 m deep) and the eggs are spiral with six to seven turns on the sides and long slender tendrils emerging from the apex of the shell case — these tendrils help secure the eggs to algae or sponges. Port Jackson Sharks deposit their eggs in much shallower water, often less than 5 m deep, and usually within rocky crevices in

◁
A Banded Wobbegong resting in a sandy gutter. Western Australia. *Barry Hutchins*

◁
An aggregation of Port Jackson Sharks at Jervis Bay, New South Wales. *Joachim Ngiam*

◁
A Spotted Wobbegong resting on sand. The white spots on its back are clear, as is the fringe of dermal flaps around the rim of the mouth. Montague Island, New South Wales. *Ken Hoppen*

sheltered bays such as Sydney Harbour and Jervis Bay. Their eggs have four to five turns on the sides and have either very small, inconspicuous tendrils, or none at all.

Port Jackson Sharks were the subject of a series of important ecological studies in New South Wales, at Sydney and Jervis Bay by O'Gower, McLaughlin and Nash during the 1960s and 1970s. They were important because they are among the first studies in Australia to make direct observations of temperate reef sharks, rather than indirectly by analysis of catch records. The scientists also tagged sharks underwater and explored short- and long-term movements of the species. Analyses of movement patterns were supplemented by protein analysis to test hypotheses about the structure of the different populations. Their studies provided good insights into the natural history, the behaviour and ecology of a reef-dwelling shark.

Port Jackson Sharks are most abundant on inshore rocky reefs in late winter and spring. Most (up to 85 per cent) of these sharks are females but small numbers of males may be found. In New South Wales, Port Jackson Sharks move large distances and completely disappear from inshore rocky reefs during autumn and winter when they move out into deeper water on the continental shelf. Adult Port Jackson Sharks have been caught by research trawlers in water as deep as 280 m off southern New South Wales. Relatively little is known about the males as they appear to spend most of their time in deeper water where they can make up 42 per cent of commercial catches.

Although often seen alone on rocky reefs, Port Jackson Sharks can been seen during the day in dense aggregations on particular reefs, returning to the same overhangs and crevices after nightly foraging. Such aggregations are particularly common in Jervis Bay where their preference for only a small number of reefs is well known. Densities of up to 20 sharks/120 m^2 have been recorded in Jervis Bay.

Females lay 10–16 eggs each on rocky reefs around October. The eggs hatch after about 12 months and the small juveniles initially forage in the bays before moving into coastal waters. Males become sexually mature when they are 8–10 years old and females when they are 11–14 years old. At the onset of winter, females move northward again, this time into deeper water on the continental shelf at around 60 m depth and up to 25 km off the coast but return to the inshore rocky reefs at the end of this migration. After laying their eggs, females begin migrating southwards along inshore reefs. Some tagged sharks have been recorded up to 850 km south of where they were tagged. The fastest rate of movement recorded was 6.5 km/day for a shark that moved 200 km in 31 days.

WOBBEGONGS, CATSHARKS, CARPET SHARKS AND EPAULETTE SHARKS

A close look at most temperate rocky reefs will often reveal the presence of a variety of small- to moderate-sized, cryptic or camouflaged sharks. These sharks seem well adapted to the reef environment and, superficially at least, most do not fit the classical mould of ferocious predators. Some are very drab in colour, others have striking disruptive colouration and divers can swim very close to them and be totally unaware of their presence. These sharks are commonly referred to as carpet sharks. Five families are represented on temperate Australian reefs: wobbegongs (Family Orectolobida), collared carpet sharks (Family Parascyllidae), blind sharks (Family Brachaeluridae), catsharks (Family Scyliorhinidae) and bamboo or epaulette sharks (Family Hemiscyllidae). With the exception of wobbegongs, these sharks are generally small (< 1 m long) weak swimmers that (foolish) divers can catch and hold by hand. Good examples of this group of sharks are catsharks, such as the Grey Spotted Catshark (*Asymbolus analis*) and the Blind Shark (*Brachaelurus waddi*), a small, cryptic grey shark with pale spots, often seen in crevices. The latter's name comes from its very small eyes which, when caught by fishers, roll back into its eye sockets, giving the appearance of being blind. Most of these sharks prey on small reef- or sand-dwelling invertebrates and small fish.

Probably the best known carpet sharks on Australian temperate reefs are the wobbegongs. Seven species occur worldwide, six of which are found in Australian waters and four of these are found on temperate reefs. Wobbegongs have a distinctive, flattened body with a series of narrow flaps of skin along the side of the head and 2–3 functional rows of needle sharp teeth. They are generally ornamental in colour, with saddles, spots and blotches. All species appear to be viviparous, giving birth to free-swimming active young.

The four species likely to be encountered on Australian temperate reefs are the Spotted Wobbegong (*Orectolobus maculatus*), the

Ornate or Banded Wobbegong (*O. ornatus*), the Western Wobbegong (*Orectolobus* sp. nov.) and the Cobbler Wobbegong (*Sutorectus tentaculatus*). The latter is relatively slender, with few dermal flaps and is readily distinguished by the presence of rows of wart-like protuberances, called tubercles, along the back. It grows to less than 1 m in length and occurs from South Australia to the Houtman Abrolhos Islands. The Western Wobbegong occurs in Western Australia from Cape Leeuwin to Coral Bay. Juveniles have two rows of small tubercles from the head back along the body, and relatively few dermal flaps. Colour is chocolate brown to yellowish on top, and ornamented with darker saddles that have indistinct margins. These sharks grow to about 2 m in length.

The two most commonly seen members of the family are the Spotted and Banded Wobbegongs. The former has a yellowish to tan back, with light, 0-shaped marks — often white-edged — which obscure darker saddles. There are about 6–10 dermal flaps below and in front of the eyes. This shark grows to at least 3 m in length and occurs around southern Australia from Moreton Bay in Queensland to Fremantle in Western Australia. The Banded Wobbegong is extremely ornate, having a yellowish brown to greyish brown back with corrugated saddles. Each saddle has paler blue to white patches. There is a prominent white spot behind each spiracle. Banded Wobbegongs grow to about 2.9 m in length and occur from Papua New Guinea down the east coast of Australia, across southern Australia and north to the Houtman Abrolhos Islands.

Wobbegong sharks appear to be masters of ambush predation, being able to rest motionless on the seabed until suitable prey passes by — then they react quickly to seize it. Surveys of Spotted Wobbegongs at Seal Rocks, New South Wales found the same sharks in the same location over three days — suggesting they move little or that they return to the same position after moving about. Wobbegongs eat a variety of reef fishes and invertebrates and they are often captured in fish traps or lobster pots — either chasing the bait or the fish and rock lobsters attracted into the traps. Spotted Wobbegongs are caught in New South Wales by long-lining over reefs. Little is known about the dynamics of this fishery but the sedentary nature and relatively few offspring produced by this species suggests that they are less likely to sustain intense fishing than many fishes.

GREY NURSE SHARKS

Sharks in the family Odontaspididae comprise four species worldwide, two of which are found in temperate Australian waters. The group is found in almost all the temperate and subtropical seas of the world. The sharks are generally known as sand tiger sharks, and in southern Africa they are called ragged-tooth sharks. The name 'grey nurse' is used only in Australia. In fact, the name 'nurse sharks' is generally applied to the Family Ginglymostomatidae, which are large, sluggish sharks normally confined to tropical and subtropical waters.

In Australia the Grey Nurse Shark is found in temperate waters on shallow inshore reefs and on the continental shelf out to depths of almost 200 m. The species is most commonly found in southern Australia, from southern Queensland to Shark Bay in Western Australia. Grey Nurse Sharks are rare in tropical Australia and do not occur in Tasmanian waters. The second species of odontaspidid shark found in Australian waters is the Smalltooth Sand Tiger or Herbst's Nurse Shark. It occurs in much deeper waters than the Grey Nurse Shark; in New South Wales it has been caught only on the slope of the continental shelf in depths between 300 m and 800 m.

Grey Nurse Sharks have a stocky body, relatively flattened head and five gill slits. The two dorsal fins lack spines and are about the same size, and the tail has a elongated upper lobe. Just before the tail (the area known as the caudal peduncle) there is a small but distinctive pit. The teeth of Grey Nurse Sharks are long, narrow and sharp and there are two small points (or cusps), one on either side of the teeth. In Herbst's Nurse Shark there are 2–3 cusps on each side of the central point of each tooth.

Grey Nurse Sharks display some unusual behaviours. Firstly, they are believed to maintain their buoyancy in the water by gulping air into their stomachs. In this way they are able to hover motionless just above the reef floor. Secondly, foetal Grey Nurse Sharks display a somewhat gruesome behaviour during pregnancy. Like many other sharks, they are viviparous, which means that the mother gives birth to live free-swimming young, usually one per uteri. The first egg to hatch in each of the two uteri eats other unfertilised eggs or smaller, newly hatched foetuses. This practice is known as uterine cannibalism or adelphophagy (meaning eating of siblings). The two pups that are born usually have a store of eggs

▷
The Grey Spotted Catshark is sometimes found in crevices and gutters in south-eastern Australia. Merimbula, New South Wales.
Ken Hoppen

▷
Groups of Grey Nurse Sharks are an imposing sight. Malua Bay, New South Wales.
Paul Baumann

△
A Circular Stingaree (*Urolophus circularis*) on sand near a reef. Geographe Bay, Western Australia. *Sue Morrison*

within their stomachs to provide sustenance until they can begin capturing prey. Larger Grey Nurse Sharks feed on a variety of fish, including other sharks and rays, and some of the large reef invertebrates such as octopuses, cuttlefish and rock lobsters. Prey is usually captured and held with the sharp, grasping teeth and swallowed whole. It is believed that most feeding occurs at night.

Sand Tiger Sharks undergo long migrations along the continental shelves in many parts of the world. In Australia, Grey Nurse Sharks, like the Port Jackson Sharks, also migrate along the east coast of Australia, and probably also along the west coast. Our understanding of these migrations comes from a variety of sources, including records from gamefishing and protective meshing of swimming beaches, and anecdotal information from divers. Grey Nurse Sharks appear to migrate south from southern Queensland along the New South Wales coast during summer and return to the north, beginning in autumn. This pattern of migration is not consistent with movements of the species found along the east coast of southern Africa. There, odontaspidid sharks gather in warm tropical waters in summer, apparently to mate. The pregnant females then travel south to give birth in winter (after a gestation period of 9–12 months). Little is known about the movements of the males.

Grey Nurse Sharks are seen commonly at Seal Rocks, on the mid-north coast of New South Wales. Here they frequent a large cave at Big Seal Rock and several rocky gullies on the western sides of Big Seal Rock and Little Seal Rock. Underwater visual surveys at Seal Rocks also indicate that Grey Nurse Sharks are common in these gullies during the morning, but some move away — possibly to feeding grounds — in the afternoon. More work, however, is required to confirm these patterns.

Grey and Herbst's Nurse Sharks have been totally protected in New South Wales waters since 1984 — the first species of sharks in the world to be given this status. Some of the ways in which we can gain a better understanding of the ecology of sharks on our temperate reefs is to do more underwater surveys, preferably at many sites and different times of the year, to document the extent and timing of migrations. Smaller-scale movements could be examined by tracking sharks over daily cycles using acoustic tags. This type of information could help to resolve the conflict between divers and fishers and help define the most appropriate management strategies to preserve the species.

STINGRAYS AND STINGAREES

Normally we associate stingrays with soft sediments and usually see them gliding gracefully over areas of sand or lying in patches of sand within rocky reefs. They feed on a variety of reef- and sand-dwelling invertebrates, and can crush shells in their powerful jaws. Anglers occasionally hook Smooth and Black Stingrays, although they are rarely landed due to their very large size. They are also taken in trawl nets. Stingrays belong to the family Dasyatidae. Stingarees (Family Urolophidae) are also relatively common but are generally much smaller than stingrays and found in deeper water. They often have long, tapering tails, armed with one or more venomous spines. These spines are usually serrated and capable of inflicting severe wounds. Stingrays and stingarees are live bearers.

Two species of stingrays commonly found on temperate Australian reefs are the Black Stingray (*Dasyatis thetidis*) and the Smooth Stingray (*D. brevicaudata*). Both species occur around the southern half of Australia and Tasmania, on the east coast of southern Africa and off New Zealand. They extend from very shallow water to depths in excess of 350 m. In some areas they are known to swim into water less than 1 m deep, foraging around boat ramps, fish-cleaning areas, etc. The Smooth Stingray has a greyish brown dorsal surface, often with dark colouration around the eyes. The tail is relatively short compared to other stingrays and the back is quite smooth. It is believed that this was the species caught by Captain Cook when he first arrived in Australia, naming the embayment 'Stingray Bay' — this name was later changed to Botany Bay. The Smooth Stingray is one of the world's largest rays, growing to lengths of more than 4.3 m, with a disk width of over 2 m and weighing in excess of 350 kg. It is believed to occur in shallow waters during summer and autumn, possibly migrating to the deeper waters of the continental shelf in the winter.

The Black Stingray, also known as the Thorntail Stingray, is much darker than the Smooth Stingray, varying from dark grey to black on its upper surface. The back has a cover of hard 'thorns' along the midline and, beyond the tail spine; these thorns cover the tail. The tail is relatively much longer than in the Smooth Stingray and can be twice the length of the disk. This species grows to a total length of 4 m, with a disk width of 1.8 m and a total weight exceeding 210 kg.

CONCLUDING REMARKS

This chapter has identified some of the characteristics about reef sharks and rays we do know, some of the attributes we do not know and some of the features that we need to know to better understand and conserve these components of temperate reefs. Sharks and rays probably have a relatively minor role in the ecology of rocky reefs and their loss may have little effect on the rest of the ecosystem. The loss of these species would, however, diminish these ecosystems in the same way that the loss of large predatory cats would diminish the African plains or the loss of large fishes would reduce our expectations of kelp forests. While we may marvel at the many and varied features of temperate rocky reefs, the sight of a large Grey Nurse Shark, ornately patterned wobbegongs or massive stingrays has the ability to enthral us in ways that few other reef animals can.

Yellowtail Kingfish school around a floating raft. Sydney, New South Wales. *John Matthews*

18 Snapper & Yellowtail Kingfish

Gary Henry and Bronwyn Gillanders

In addition to the species of fish that live permanently on rocky reefs, there are many species that are only occasional visitors. Of these fish, probably best known are Snapper (*Pagrus auratus*) and Yellowtail Kingfish (*Seriola lalandi*). Both species are found in a wide variety of habitats, including the open ocean, on the continental shelf and in estuaries. Snapper and Yellowtail Kingfish (or more simply Kingfish) find a place in this book because of their high profile with anglers, spearfishers and scuba divers on rocky reefs.

SNAPPER

Snapper is an attractive species with red–pink colouration on the head and upper body, paler sides with scattered bright blue spots, and a creamy white belly. Large Snapper often develop a prominent hump on the top of their head, and the snout may become fleshy and pronounced. These specimens are often called 'old-man' Snapper, but since many of them are female, this term is misleading. Snapper have a number of common names (eg, Pink Snapper, Cockney Bream, Squire, Red Bream, Reddies, Pinkies, Knobbies) which refer to particular stages of their life cycle. Indeed, the name Snapper is a misnomer as applied to this species because of its similarity in appearance to the true snappers (Family Lutjanidae).

Snapper are widely distributed in the subtropical and temperate continental shelf waters of the Indo-West Pacific. The species has been the subject of considerable research because of their great value to commercial and recreational fisheries in Australia, New Zealand and Japan. Snapper are somewhat unusual in that they occur in both the northern and southern hemispheres; the species is found in China, Japan, northern Indonesia, southern Australia and New Zealand but not in the intervening tropical waters. Southern-hemisphere fish are generally referred to as Snapper, while those from the northern hemisphere are called Red Sea Bream.

△ A large Snapper. *Neville Coleman*

This unusual distribution has been attributed to a relatively recent dispersal of fish across the equator, possibly during the first Quaternary glacial period. In both the northern and southern hemispheres, Snapper are confined to warm subtropical and temperate waters between latitudes 20° and 46°. Water temperature is considered to be the most important determinant of these broad patterns of distribution.

Juvenile Snapper are abundant in estuaries, on rubble and rocky reefs near the coast and on the continental shelf. Coastal rocky reefs play an important role in the life history of Snapper. Juveniles may move from flat sandy areas to rocky reef and rubble areas as they mature. The demands for space and competition may precipitate the movement of Snapper from the estuary to the coastal reefs, while the effect of predation may slow this movement within estuaries. Adult Snapper move into deeper water and are found in shelf waters between 20 m and 60 m depth, but their distribution ranges down to 200 m, usually over reef. The fragmented nature of the commercial and recreational fisheries for Snapper indicate that the distribution is not uniform and that aggregations of Snapper occur in preferred localities.

AGE AND GROWTH

Age and growth of Snapper have been studied in Australia, New Zealand and Japan using a variety of techniques. In eastern Australia, Snapper appear in coastal waters and marine estuaries at a length of approximately 20–30 mm and an estimated age of two months. Growth appears to be rapid at first, but the rate of growth declines after maturity and the value of successive yearly-size increases progressively decline. The mean lengths of 1-, 2-, 3-, 4- and 5-year-old Snapper from Coffs Harbour were estimated to be 185 mm, 230 mm, 272 mm, 311 mm and 348 mm respectively. Maturity is attained during the second or third years and Snapper grow to a maximum size of about 1 m over a period of 35–40 years. In New Zealand, Snapper may live to 60 years of age.

Snapper longevity, growth and maximum size vary with region, season, maturity and habitat. Water temperature is considered to be the most important influence on Snapper growth because summer and autumn are periods of rapid fish growth, and winter and spring are seasons of negligible growth. Differences in food availability and competition also cause variation in growth and age.

REPRODUCTION

The reproductive biology of Snapper is broadly similar throughout its Australasian range but the timing of spawning may vary regionally. In eastern Australia, Snapper spawn in winter off Queensland, in spring off New South Wales and in summer off Victoria. This apparent correlation between spawning season and latitude has been observed for populations throughout Australasia. The environmental cues that initiate the development of eggs and sperm are considered to be a mixture of daylength, water temperature and food availability. Of these, water temperature has been nominated as the most important environmental determinant of Snapper reproductive development. Spawning, hatching of eggs and optimal larval survival generally require water temperatures in excess of 18°C.

All snapper begin life as females, but before they become sexually mature at 2–3 years old and about 300 mm long, about half the population changes sex to become males before they reproduce. Scientists have found both male and female reproductive tissue in fish that are in the process of changing sex but, because this happens before they mature, Snapper only ever function as males or females, but not both. The factors causing sex change are poorly understood.

In northern New South Wales, the season of reproductive activity extends from July to September with the major activity occurring during August. Snapper spawn repeatedly over this period. Batches of eggs and sperm are broadcast into the water over this three-month spawning season. The females spawn, on average, about twice per week during the spawning season. The number of eggs produced by each female is proportional to body size, so the annual egg production of individual fish varies widely. Each batch contains between 15 000 and 124 000 eggs, giving an annual egg production of 300 000–4 000 000 eggs per female. The timing of a species' reproductive cycle is a compromise involving many environmental considerations. Prevailing conditions of temperature, light, nutrients and therefore food availability ultimately determine the survival of the young. Snapper eggs are approximately 1 mm in diameter and float close to the surface for several days before hatching. Larval Snapper drift with the prevailing currents for 2–4 weeks before settling out of the water column onto the bottom.

MOVEMENT

Information from tagging studies indicate that Snapper do not move as much as many coastal marine fish. Most tagged Snapper are generally recovered from waters within 50 km of their release location, although a few fish may move larger distances. Patterns in the movement of Snapper suggest that individual fish move in a haphazard fashion rather than the directed migration common in fishes such as Eastern Australian salmon (genus *Arripis*) (marketing name: Australian Salmon), Sea Mullet (*Mugil cephalus*) or Tailor (*Pomatomus saltator*) which are known to migrate several hundred kilometres along the coast. Apart from tagging experiments, studies of the genetic fingerprint of Snapper, trace metal residues in muscles and bony parts, minor differences in the shape of the fish and the degree of infestation with parasites, confirm that Snapper is a relatively sedentary species.

YELLOWTAIL KINGFISH

Yellowtail Kingfish is a wide-ranging pelagic species of both commercial and recreational importance. Scuba divers may notice a school of Kingfish suddenly appear, inquisitively circle them and then disappear. They are easily recognised by their streamlined, elongate body with a yellowish stripe that runs through the eye and along the length of the body. Despite its popularity, relatively little is known of the life history or status of the Kingfish populations.

DISTRIBUTION AND FEEDING

Kingfish are members of the family Carangidae, which includes the jacks and trevallies. This family has about 140 species including Samson Fish (*Seriola hippos*), Jack Mackerel (*Trachurus declivis*) and the trevallies (in the genera *Caranx* and *Carangoides*). Within the genus *Seriola*, there are nine species, of which four are found in Australian waters: Yellowtail Kingfish, Amberjack (*S. dumerili*), Highfin Amberjack (*S. rivoliana*) and Samson Fish. Kingfish are widely distributed in the cool-temperate waters of the Pacific and Indian Oceans and are abundant in Australia, New Zealand, South Africa, India, Japan, Hawaii and California. The Australian species of Yellowtail Kingfish is thought to be physically similar to populations off California and Asia, but because of their geographical separation they may not interact with these populations and are therefore regarded as a separate subspecies.

△
A young snapper showing the irridescent blue spots that make this species so attractive.
Mary Malloy

In Australia, Kingfish are found from Queensland, around the southern coast of the continent to Western Australia. They also occur on the east coast of Tasmania and around Lord Howe and Norfolk Islands. Kingfish are highly mobile and form aggregations over seamounts and inshore reefs, around rocky headlands, and in oceanic waters around drifting objects such as seaweed and buoys, which provide structure in an otherwise featureless world, concentrating prey and offering shelter from larger predators.

Kingfish mainly feed on small schooling fish, although squid and crustaceans are also eaten. The remains of these prey are often identified by hard parts (for example squid beaks) that remain in the guts of Kingfish long after the softer parts have been digested. Kingfish are thought to feed during the day rather than at night.

AGE AND GROWTH

The largest Kingfish reported in Australasia was heavier than 60 kg and more than 2 m long. As with many species of mobile schooling fish such as tuna, Kingfish are difficult to age using the conventional methods of counting rings in hard parts such as ear bones (otoliths), scales and vertebrae. Their otoliths are very small compared to the actual size of the fish, for example a 500 mm long Kingfish has otoliths that are only 5 mm in length. This compares with Snapper in which a 300 mm fish has otoliths that are 12 mm long. Thin cross-sections of the otoliths of Kingfish show numerous striations, which can rarely be interpreted as annual growth zones. These difficulties notwithstanding, estimates of growth in Kingfish suggest that there is an initial period of rapid development followed by progressively slower growth as the fish matures. Kingfish in New South Wales attain lengths of about 500 mm, 650 mm, 750 mm and 850 mm in the first 4 years of life respectively. These estimates of age have been corroborated using other methods such as measuring changes in size of tagged fish.

REPRODUCTION

There have been few detailed studies of the reproductive biology of Kingfish. While some reproductive condition may be apparent in Kingfish during all months of the year, peaks in spawning activity are observed in the spring or summer months. It is unknown whether the timing of spawning is uniform over the known range of Kingfish in Australia, but along the Californian coast, spawning occurs at approximately the same time in all areas. Spawning of Kingfish may be modified by coastal water temperatures, although other physical and chemical characteristics of the marine environment are likely to be involved.

Sexual maturity in Kingfish is thought to occur around 3–4 years and at lengths of 550–850 mm. Kingfish in New South Wales are thought to mature towards the upper end of this size range, although more research is needed to confirm this observation. Males appear to mature at smaller sizes and younger ages than females. Kingfish eggs are pelagic and 1–1.4 mm in diameter. They hatch into larvae within 2–3 days at a size of 2.7–3.8 mm. Juvenile body characters and pigmentation are attained by a size between 9 mm and 19.4 mm. Small juveniles are very different from adults in terms of colouration. They are yellowish with dark brown vertical bands that gradually fade as the fish increases in size. Little is known of the life history of Kingfish from larval stages to juvenile recruitment to coastal waters.

MOVEMENTS

Most recoveries of tagged Kingfish are within 50 km of where the fish were released. However, a small proportion of Kingfish move a considerable distance from their release location and several trans-Tasman migrations (>2000 km) have been recorded including fish moving in both directions. There are also records of fish moving in excess of 500 km along the east coast of Australia and between New South Wales and Lord Howe Island. Kingfish are therefore capable of extensive movements, but whether these movements are related to spawning or feeding migrations is unknown. There is some evidence to suggest that the long-distance movements are undertaken by adult Kingfish (>750 mm) while juvenile fish remain in a relatively discrete area.

A Bicolor Scalyfin.
Lucky Bay,
Western Australia.
Barry Hutchins

19 Territorial damselfishes

Michael Kingsford

Damselfishes are found in all the oceans of the world, but are most abundant and diverse in the tropics. About 110 species of damselfish are found in the waters around Indonesia and perhaps 80 species are recorded in tropical Australia. With the exception of the subpolar regions, damselfishes are found wherever there are rocky reefs. In temperate regions of Australia their diversity drops rapidly with increasing latitude but damselfishes are still a conspicuous component of the local fish fauna on reefs in the southern states. In Western Australia and New South Wales tropical damselfishes such as Gold-belly Gregories (*Stegastes gascoynei*), Pacific Gregories (*Stegastes fasciolatus*) and Miller's Damselfish (*Pomacentrus milleri*) may be seen in late summer and autumn. These fish have been brought south by the Leeuwin and East Australian Currents as larvae and typically do not survive their first temperate winter.

Broadly, damselfishes either live their adult lives close to the reef in territories or are planktivorous and are seen high in the water over reefs, often in schools (see Chapter 25). Of the territorial damselfishes, scalyfins (genus *Parma*) are the largest and often most abundant. Scalyfins will be familiar to everyone who has dived on rocky reefs in southern Australia. Although these fish are rarely longer than 25 cm, they stoutly defend their territories against all comers, even divers. Although some scalyfins are found in tropical waters, the group is essentially temperate in its distribution and most species are found only in southern Australia.

In most parts of temperate Australia, at least one species of the group are abundant. In New South Wales and southern Queensland, Big-scaled Parma (*Parma oligolepis*) and Banded Parma (*P. polylepis*) are most abundant. In northern New South Wales, Girdled Scalyfin (*P. unifasciatus*) predominate in the *Pyura* Habitat (see Chapter 2) and further

◁
A Western Scalyfin. Houtman Abrolhos Islands. Western Australia. *Gerry Allen*

◁
A Victorian Scalyfin. Recherche Archipelago, Western Australia. *Gerry Allen*

◁
A male White Ear in its territory. Sydney, New South Wales. *Rudie Kuiter*

△
A juvenile Victorian Scalyfin. Portsea, Victoria. *Rudie Kuiter*

south they overlap with White Ears (*Parma microlepis*). White Ears are generally the most common species in central and southern New South Wales but are found as far south as Tasmania. In Tasmania, and from Wilsons Promontory in Victoria to Western Australia, Victorian Scalyfins (*P. victoriae*) are the most abundant territorial damselfish. The Bicolor Scalyfin (*P. bicolor*), Western Scalyfin (*P. occidentalis*) and McCulloch's Scalyfin (*P. mccullochi*) are found only in Western Australia.

Most species of scalyfin are drab in colour as adults but brightly coloured as juveniles. Male White Ears are black and the females are brown, although this colour difference is relatively subtle. Adult Banded Parma and Big-scaled Parma from northern New South Wales are slightly more colourful, as are Girdled Scalyfin and the white and brown Victorian Scalyfins. Adults of the West Australian Bicolor Scalyfin are gaudy by comparison to most scalyfins. The colour differences between adults and juveniles undoubtedly caused some confusion with early taxonomists who probably thought the juveniles were separate species. It is unknown why these colour differences occur, but it may relate to mating behaviour as brightly coloured non-reproductive juveniles are not subjected to the strong aggression observed among reproductive fish.

CLOWN FISHES

Although clown fishes are some of the best-known tropical reef fishes, members of the genus *Amphiprion* are found in subtropical reefs in the same close association with anemones. Clown fishes are found on the east and west coasts of Australia, at the Solitary and Houtman Abrolhos Islands. Clown fishes often have a specific relationship with the host anemone. The Blue-lip Anemonefish (*Amphiprion latezonatus*) is found in southern Queensland and northern New South Wales and is only found in the anemone *Heteractis crispa.* McCulloch's Anemonefish (*Amphiprion mccullochi*) is only found in the Lord Howe Island group in the anemone *Entacmaea quadricolor.* The Australian Anemonefish (*Amphiprion rubrocinctus*) is found in Western Australia and is hosted by two species of anemone (*Entacmaea quadricolor* and *Stichodactyla gigantea*).

Clown fishes are not only known for their startling colours and association with anemones, but also because they change sex and typically form monogamous pairs in which the female is always the larger of the two. The female of the pair first reproduces as a male before changing sex into a female. Within a single anemone there are usually adults and juvenile fishes. The juvenile fishes in the anemone are aggressively inhibited from reproducing until one of the adults disappears. When the female is removed, the male quickly changes sex into a female and the largest juvenile matures to take over the male role (for more on sex change see Chapter 21).

A juvenile Blue-lip Anemonefish. South West Rocks, New South Wales. *Ken Hoppen*

WHERE DO YOU FIND SCALYFINS ON REEFS?

Scalyfins are found in a wide range of habitats including the mix of corals and temperate algae found at the Solitary and Houtman Abrolhos Islands, on reefs dominated by Cunjevoi (*Pyura stolonifera*) in northern New South Wales, urchin-grazed Barrens Habitat in central and southern New South Wales and the reefs covered with large brown algae that are typical of Victorian rocky reefs. Individual species can exhibit strong habitat preferences within a latitudinal range. Although White Ears are found at all depths, in central New South Wales they are most abundant in urchin-grazed Barrens Habitat and relatively uncommon in kelp forests. In the Barrens Habitat, White Ears can be found at densities

A female Victorian Scalyfin laying eggs, attended by the darker-coloured male. Popes Eye, Queenscliff, Victoria. *Geoff Jones*

HOW DO YOU AGE A FISH?

Fish can be aged by using marks or rings laid down in their bones, much like rings that are used to age trees. Daily and annual rings are formed in fish. Daily rings are usually used to age larval and juvenile fishes in their first year of life. Annual rings are found in many fish and often form at times of the year when the fish are growing slowly. The body temperature of most fish is similar to that of the surrounding seawater and temperature has a great influence on the growth of fish. Growth is usually slowest in winter. Bones that are commonly used to age fish are ear bones, cheek bones, spines and vertebrae. Some reef fish live to be older than 50 years and novel methods such as chemical analyses have been used to age them. In Snapper (*Pagrus auratus)* aged using these techniques, fish older than 50 have low levels of radioactive carbon near the centre of their otoliths because they were alive during the period from 1945 when nuclear weapons were exploded in the Pacific region.

of up to 20/100 m^2, but more typically at densities of 10–14/100 m^2. The brightly coloured juveniles are found in greatest numbers in summer and autumn and, during this period, they are most common in the Barrens Habitat in water less than 10 m deep.

During the breeding season, a female will leave her own territory and enter the territory of a male to lay her eggs which are laid in a cluster on a vertical rock surface that has been cleared by the male. The female and the male alternate in laying a small batch of eggs and fertilising them.

Victorian Scalyfins are found on reefs with a high cover of the large brown algae that are typical of Victorian rocky reefs. Victorian Scalyfins can be found in densities of up to 5/100 m^2 but more typically at smaller densities of 1–4/100 m^2. Many of these habitat preferences probably relate to the dietary needs of the fish, shelter (crevices are abundant in Barrens Habitat) and their requirement to lay eggs on the bottom, which are subsequently guarded by the males.

TERRITORIALITY

Scalyfins are strongly territorial and, in New South Wales, female White Ears do not normally stray from their territories except to chase intruders (of their own or other species), often well outside their normal range before giving up the chase. During the spawning season, females will venture well outside their territories to assess the 'suitability' of nearby males. Males and females are found over the same depth range (usually 3–25 m) and members of each sex are generally found close to crevices and boulders that offer shelter. A paradox for herbivorous White Ears is that they are most abundant in areas with little algae. This suggests, perhaps, that other needs such as shelter and reproduction are more important than the availability of large brown algae. Victorian Scalyfins are also strongly territorial and defend territories of between 3 m^2 and 30 m^2. The size of their territories is determined by the local densities of fish. When fish are removed, the territories of remaining fish expand to fill the gaps.

AGE AND GROWTH

Scalyfins are relatively small fish but they can grow to be quite old. For example, most adult White Ears are 10–20 years old and the oldest fish recorded in New South Wales was 37 years old. The majority of fish with juvenile colouration are less than two years old, although a few five-year-olds may retain their juvenile colouration. In their first five years, fish grow to be between 170 mm and 175 mm long. It is difficult, therefore, to tell the difference between a 10-year-old and a 30-year-old White Ear. All males and females older than six years are sexually mature and the youngest age at maturity recorded is three-years-old.

REPRODUCTION

In contrast to the wrasses (Chapter 21), sex change is not common in damselfishes. Sex change does occur in some tropical species of damselfish (eg, the Humbug Damselfish, *Dascyllus aruanus*) but this is thought not to be the case for temperate damselfishes, although data exist only for White Ears. More information is required on general patterns of sexuality in these fishes. The spawning behaviour of scalyfins is easy to observe because they lay their eggs directly on to the reef within clearly defined territories rather than broadcasting their eggs and sperm into the water. White Ears in New South Wales lay their eggs between October and January. Unlike many damselfishes, which spawn only once per season, White Ears spawn in bouts as their eggs become mature and, as a result, eggs within and among nests are at different stages of development.

Female White Ears leave their territories to spawn in the nests of the males. Males court females at all times of the day. When a female approaches the nest of a male he displays by swimming around her quickly in small circles. The male then attempts to lure the female back to his nest by waggling his tail repeatedly. Audible clicks and grunts can often be heard from the male when the female enters the nest. The female lays her eggs onto the substratum and these are fertilised by the male. The spawning success of males varies greatly, from those that rear no eggs, to those that guard up to 8000 cm^2 in a season (each egg is about 1 mm in diameter). Male size, aggression and habitat complexity does not explain variation in reproductive success, but the quality of the nesting site within a territory may have some influence on the number of eggs a male fertilises. Males guard the eggs until they hatch about 12 days later. The nests can be observed during the spawning season, usually under ledges, and within the nests the eggs may be seen. Individual fertilised eggs are easy to see during the latter stages of development as the embryos have conspicuous silver eyes. The larvae usually hatch out at night and spend up to a month in the plankton before they return to rocky reefs as juveniles at about 12 mm long.

WHAT DO THEY FEED ON?

Territorial damselfishes mostly consume algae, but animal material has been found in the guts of many fish. Although plant material is dominant in terms of volume, there are no data on how important the animal components of the diet are to the nutrition of damselfishes. In New South Wales, White Ears can be seen actively pecking at apparently barren rock surfaces, on the surface of kelp plants and in the water column. Fish of all sizes have a high percentage of plant material (algae) in their guts, although up to 20 per cent of the diet of White Ears is animal material such as sea squirts, plankton and little crustaceans. Small White Ears have a much higher percentage of animal material in their guts, as is common for fish that are herbivorous as adults. It is unlikely that feeding by these fish has a significant influence on the distribution and abundance of algae, particularly in comparison with the sea urchins and limpets that are abundant in habitats where White Ears are also abundant. Densities of territorial damselfishes may influence the nature of the diet. For example, it has been found that Victorian Scalyfins consume a wide range of algae, but some types are preferred (eg, the epiphytic red algae, *Champia* and *Polysiphonia*). Fish with large territories generally obtain more preferrred food. Although Victorian scalyfins are capable of removing kelp from their territories, it is unknown whether this is a frequent behaviour.

WHO DO THEY INTERACT WITH?

Little is known of the predators of temperate damselfishes, but small fish are most vulnerable to predation and are probably lost to a host of predators such as Half-banded Seaperch (*Hypoplectrodes muccullochi*) and vagrant fish such as large trevally. Damselfish of all sizes may also be taken by gropers, beardies, cuttlefish, moray eels (genus *Gymnothorax*) and sharks such as wobbegongs.

Scalyfins interact with most organisms on reefs and I have observed many different species of territorial damselfishes chase there own kind and a diversity of other species. They are particularly aggressive toward herbivorous fishes such as Rock Cale (*Crinodus lophodon*) that may consume the algae in their territories, but other intruders that are not herbivorous, such as morwong, bream and wrasses, are also attacked. Even divers are not immune from the aggression of these tenacious little fishes.

Red Morwong showing their large eyes and small mouths. Bermagui, New South Wales. *Rudie Kuiter*

20 Morwongs

Michael Lowry and Mike Cappo

Morwongs (Family Cheilodactylidae) are an important and familiar component of the fish fauna of temperate rocky reefs in Australia, New Zealand, South Africa, Japan and, to a lesser extent, South America. Morwongs are popular with underwater photographers and spearfishers because they are easy to approach and have striking colour patterns. Some deeper-water species also support important trawl and line fisheries.

DIVERSITY AND SIZE

The cooler waters of Australia and New Zealand have the greatest diversity of morwongs in the world — three of the four genera in the family and 11 of the 20 species (55 per cent of the entire family). Ten species of morwong are found only in Australasia. The taxonomy of the family is presently under review, however, and the 20 species presently known may be increased to nearly 30 as the group is better understood. There are several characteristics of the group that will be familiar to many divers and anglers. Most obvious are the enlarged pectoral fins in which 4–7 of the lower fin rays are extended, separate and fleshy. In fact, the scientific name for Family Cheilodactylidae means 'lips and fingers' and the fish are called 'finger-fins' in South Africa. These extensions were once thought to be used as feelers, but underwater observations have shown that they are used to stabilise or pivot the fish as it rests or feeds on the reef, and possibly to preen or clean the tail and body. Morwongs have large eyes and a small terminal mouth with tiny peg-like teeth and large fleshy lips. All species have a pelagic, early juvenile stage known as the 'paperfish' (see page 176), which is rarely seen and changes dramatically in shape, colour and habits after settlement to the reef. The closest relatives to the morwongs are perhaps the hawkfishes (Family Cirrhitidae), sea carps (Family Aplodactylidae) and trumpeters (Family Latrididae).

The morwongs can be divided into the deep-bodied *Nemadactylus* species inhabiting deeper waters on or off reefs (and caught by trawl, trap and line fisheries), and those species associated with reefs in water shallower than 60 m that do not take a hook. The Jackass Morwong (*Nemadactylus macropterus*) is the only species found in the Indian and Atlantic Oceans, as well as southern and southeastern Australia and New Zealand. It reaches about 66 cm and 2.9 kg in Australia, schools in deep water (to 400 m), and is mainly caught away from reefs over sand. Jackass Morwong have been trawled off south-eastern Australia since 1915, together with the lesser known Blue Morwong (*Nemadactylus douglasii*), with annual catches up to 2100 t. The Blue Morwong also forms schools and reaches 81 cm and 5.9 kg. It is also known as 'Blubberlip' by charter boat anglers who fish by drifting over offshore reefs. The attractive Queen Snapper (*Nemadactylus valenciennesi*; marketed as Blue Morwong) may be more solitary, inhabiting deeper reefs to at least 240 m. It grows to 90 cm and at least 11 kg in weight, and is taken by line and as a bycatch of the bottom-set gillnet fishery for edible sharks. It is the most colourful of the *Nemadactylus* species, with an iridescent blue base colour, bright yellow stripes radiating around the eyes and yellow stripes on the juveniles.

Adults of these three species are seen by divers on deeper coastal reefs, but far more commonly seen, and more colourful, are the eight species associated with shallow reefs. The Dusky Morwong (*Dactylophora nigricans*) or Strong Fish is the largest family member in Australia (up to 1.2 m and 13.6 kg) and is commonly found inhabiting limestone reefs and seagrass beds (*Posidonia*). The other seven species can be separated into two groups by their distinctive markings and size.

◁ Banded Morwong. Merimbula, New South Wales. *Rudie Kuiter.*

◁ Magpie Perch showing distinctive black and white banding. Sorrento, Victoria. *Paul Baumann*

△
A large aggregation of Red Morwong. South West Rocks, New South Wales. *Ken Hoppen*

A group of larger species with rusty red base colours or red and white markings includes the Banded Morwong (*Cheilodactylus spectabilis*), the Red Morwong (*Cheilodactylus fuscus*), the Painted Morwong (*Cheilodactylus ephippium*) and the strikingly banded Red-lipped Morwong (*Cheilodactylus rubrolabiatus*). These species all grow to be 65–70 cm in length and 3.5–4.3 kg in weight in subtropical and temperate Australia. In New Zealand, the Banded Morwong is reported to grow to 1 m and 15 kg. By virtue of their size and sluggish habits they are popular targets for spearfishers, and the Banded Morwong has recently become the focus of an intensive gillnet fishery in Tasmania for the trade in live fish.

The second group of smaller species includes the Magpie Morwong (*Cheilodactylus vestitus*), the Magpie Perch (*Cheilodactylus nigripes*) and the Crested Morwong (*Cheilodactylus gibbosus*), all recognised by distinctive black and white banding and a maximum size less than 40 cm and about 1 kg. Although they are endemic to opposite sides of Australia, the Magpie Morwong on the eastern coast and the Crested Morwong from Western Australia both have high spines on the first dorsal fin with diagonal black stripes along the length of the pale body, giving the peculiar impression at first glance of a triangle. Magpie Perch can be identified by broad, vertical black bands, with a reddish tail. Unlike their rusty red cousins, which are often observed in aggregations, the 'magpie' varieties are frequently observed to be solitary or moving slowly over the border of reefs in pairs.

The geographic ranges of most of these species are mainly restricted to two or three mainland states, but they may be locally very common. Oceanic transport of the paperfish stage results in 'vagrants' appearing outside their common range. For example, the Painted Morwong (*Cheilodactylus ephippium*) is abundant at Lord Howe Island and frequently seen on New Zealand's offshore islands, but is rare in New South Wales.

Knowledge of morwong ecology has come from a mix of underwater observations, mostly of a small number of reef species, and fisheries research on trawled species. Morwongs of various species can be found in a variety of habitats over hard substrata from surge zones and tide-pools to seagrass beds, pier pilings and artificial reefs — although juvenile Jackass Morwong enter harbours, bays, inlets and lower estuaries, often over soft sediments. Reef morwongs prefer topographically complex habitats with a lot of shelter, such as large boulders and crevices, and are relatively inactive for much of the day. At these times they can be seen lying on their sides in crevices or using their long pectoral fins to stabilise themselves on top of suitable vantage points. Most reef species are common in water shallower than 30 m. To date only Banded and Red Morwongs and Magpie Perch have been studied in any detail and, although some generalisations can be drawn about their ecology, large gaps in our knowledge remain.

EARLY LIFE HISTORY

The early life history of morwongs has some similarities to rock lobsters in that they share a very long larval life, with the same mysteries of how they return from offshore to coastal rocky reefs. Morwong spawn small pelagic eggs that are approximately 1 mm in diameter. Larvae hatch at about 3 mm in length, quickly adopt a surface mode of life and move off the continental shelf into the open ocean. Larvae of the Banded Morwong and the Jackass Morwong have been found several hundreds of kilometres offshore. During this long pelagic stage larvae change shape profoundly as they develop; they become laterally compressed, deep bodied and almost tear-drop shaped in some species. They also develop a sharp keel along their bellies which becomes silvery below the lateral line and cobalt blue above. These adaptations make them ideally suited to life at the surface in the open ocean and probably help them avoid predators during their 9–12 month period offshore. The larger of these bizarre pelagic stages are called paperfish. So different are they from the adults in both form

▽ A 'paperfish' — the larval form of a Jackass Morwong. *Barry Bruce*

◁
Jackass Morwong form the basis of an important fishery.
Sorrento, Victoria.
Rudie Kuiter

and habits that they were initially identified as the juveniles of other types of fish.

Paperfish grow to be 45–90 mm long in the later stages of development and are fast and elusive swimmers. They are believed to combine their good swimming ability with favourable currents to return to coastal waters. But why some morwong, which are so restricted in their juvenile and adult distribution, have an early life history strategy that allows them to disperse widely is unknown. That some do disperse is evident in the capture of vagrants outside their normal range but, clearly, many still make it back to where they live their adult lives. They are cryptic at settlement and seldom photographed, but reef species are known to settle in a wide range of habitats and depths. There is a rapid and profound change into the adult form after settlement that leaves a permanent mark on scales and otoliths (ear bones).

BEHAVIOUR AND LOCAL DISTRIBUTION

Initially, small juveniles occupy exclusive 'home ranges' that they aggressively defend against intrusion by other small morwong of the same species and size by chasing them and nipping their tails. These home ranges may be as small as 26 m^2 for Magpie Perch and 100 m^2 for Banded Morwong and may include a shelter site as well as patches of reef with small turfing algae.

As the fish become larger their home ranges increase greatly in size, up to 50 000–70 000 m^2 for adult Banded Morwong. It is, presumably, impossible to defend such large areas for exclusive foraging and the fish become gregarious, forming small groups or larger aggregations with overlapping home ranges. Although densities of reef morwongs are greatest in certain favourable habitats, such as boulder banks, there is a further partitioning of the reef caused by complex social organisation in aggregations.

These aggregations are perhaps most important and best known for the Red Morwong in which activity of individuals is focused around a 'home' aggregation in fixed locations with some seasonal movement. Aggregations of females and juveniles occupy the shallow areas of the reef and larger males dominate aggregations in the deeper areas. This segregation by depth and size is similar for the Banded Morwong and Magpie Perch, but aggregations are more temporary, looser and smaller for those species.

Larger, dominant Red Morwong occupy the larger deeper crevices and the smaller fish in the hierarchy inhabit the peripheral shelters. Particular sites are chosen by the aggregations, even in areas where there appears to be no shortage of similar habitat. In late spring and early summer, local aggregations disband and converge on particular sites; the numbers of morwongs at these sites may rise from 20 or 30 to 100 fish. The factors that determine the location and dynamics of these aggregations are not understood, but may include seasonal variation in food abundance. For example, the home ranges of adult Banded Morwong are known to become larger in winter. The observed differences in the size and sex of fish occupying each aggregation are probably the result of complex interactions of behavioural factors and changes in requirements for food, shelter and breeding partners. Tagged Red Morwong that have been displaced up to 2 km from their home aggregation have returned over a period of days, passing aggregations of the same species, to the same rock they were captured from. The underlying basis for such homing ability is unknown, but the behaviour must play an important role in maintaining the observed patterns of abundance.

Marked changes in body colour may be used by the reef morwongs in maintenance of aggregations. Red Morwong can develop a series of pale vertical stripes and the Magpie Perch can alternate the position of the black and white bands in a set of display patterns as adults circle one another at close range. Blocking, chasing and nipping by male Banded Morwong have also been observed at breeding time during territorial encounters between males or when females try to leave spawning sites.

REPRODUCTION

The reef morwongs of the genus *Cheilodactylus* have distinctive orbital tubercles or 'horns' when they are mature, which protrude from the face just below the eye and provide a ready means of identifying the sex of Banded and Red Morwongs, and perhaps others. Male Red and Banded Morwongs are larger and have more pronounced horns than the females and they also have a smaller eye diameter, larger pectoral fins and shorter heads.

Studies of these species show that while larger males inhabit deeper reef habitats, ripe females from shallower areas associate with them to spawn numerous times during four to five months — over autumn and winter for Red Morwongs and summer–autumn for Banded Morwongs. The sexually mature fish do not all become ripe at the same time in these extended spawning periods and, in the case of Banded Morwongs, males typically have small testes which may be less than one hundredth the mass of the ovaries of the female. During the spawning season male Banded Morwongs become territorial and defend an area that includes several caves. The number of ripe females attracted down the reefs to visit a particular territory is related to the presence of conspicuous caves, which probably provide spawning sites around dusk. In contrast, Red Morwong form large spawning aggregations of 80–100 fish in a pyramid formation with males and females releasing eggs and sperm as they move up through the water column.

FEEDING ECOLOGY

Morwongs have a characteristic feeding pattern, eating mainly amphipods, small crustacea, sea worms and shellfish, and can be labelled as 'benthic micro-carnivores' — or perhaps 'benthic carnivores' in the case of adults because their prey get larger and the variety of prey items taken by the fish increases as they grow. The thick, fleshy lips are used to suck and wrench prey from algal turfs, foliose algae and pockets of sediment in crevices. During feeding the fish are oriented head down and the long finger-like extensions to the fins are used to both manoeuvre the fish and to act as pivot points as they peck at the reef. The 'suck and sort' feeding mode of the morwongs is usually directed at appropriate feeding sites, not visible prey, unless worms or other large prey are uncovered. The mouth cavity expands up to seven-fold during suction, so quickly that a click can be heard underwater, and the fine peg-like teeth rake the substrata. The ingested material is sorted very efficiently, with larger debris ejected through the mouth and finer sediment winnowed out through the gills.

Small amphipods generally dominate the diet of the Red and Banded Morwongs and Magpie Perch, but as the fish grow, the increased mouth gape and the suction it affords, enables them to exploit more complex topography — such as algal holdfasts — where prey such as brittle stars, sea worms, small crabs and abalone are common and can be removed by the bigger fish. The relative contribution of all these prey therefore changes as the fish grow, and in the case of the Banded Morwong the change in diet occurs at the same time as the onset of sexual maturity and a slowing down of growth when they are about 250 mm long. A similar change in diet of the Red Morwong occurs when the fish are about 200 mm long.

The smaller fish forage mainly on algal turfs and flatter surfaces and for much longer times during the day than the adults, presumably because they must harvest smaller prey. Juvenile Red Morwong and Magpie Perch also feed throughout the day. In contrast to the other species known to shelter or sleep in tight crevices at night, the adult Red Morwong are active nocturnal foragers and are relatively inactive during the day. The seasonal shifts in the location of Red Morwong aggregations may de due partly to changes in food availability, since they have a much higher proportion of amphipods in their summer diet, but field studies are lacking.

GROWTH AND AGE

Many temperate-reef species are now known to be surprisingly long lived. The morwongs are no exception, with female Red Morwong living at least 40 years and female Banded Morwong living to be older than 59 years. It is likely that some Banded Morwong live even longer because the fish estimated to be 59 years old were not near the largest fish recorded for the species. Females are known to live longer than males and, in the case of Jackass Morwong, grow faster. New Zealand morwongs may live the longest; Jackass Morwong 50 years old have been recorded there, compared to a longevity of 16 years in Australia. Causes of mortality are unknown, although the Southern Calamari Squid (*Sepioteuthis australis*) has been seen preying on juvenile Magpie Perch.

CONCLUSION

The available information indicates some similarities in the ecology of the reef-based morwongs, particularly site attachment and a progressive partitioning of the reef habitat by depth. In the case of the long-lived species that form aggregations on reefs, there may be a 'storage effect' where local populations have taken many decades to build up in number and size from local recruitment of juveniles. These traits have made the Red Morwong an ideal indicator species for monitoring of contamination from Sydney's ocean effluent outfalls, but they also mean

▷ A Red-lipped Morwong at Rottnest Island, Western Australia. *Sue Morrison*

that exploitation has the potential to severly deplete populations.

Indeed, intensive spearfishing has resulted in the depletion of larger fish and overall numbers in local populations in some areas. In New Zealand, populations of Banded Morwong in Marine Protected Areas are now significantly larger than in surrounding areas. Provision of these areas may be the best conservation method in the face of increasing exploitation, but more studies will be needed to design these areas at appropriate scales. It is also hoped that an awareness of the age and complex community organisation of morwongs associated with reefs, combined with a growing appreciation for the family among the diving community, will safeguard local populations from over-exploitation.

Crimson-banded Wrasse are common on coastal reefs near Forster, New South Wales.
Ken Hoppen

21 The wrasses

Geoff Jones

DIVERSITY

Rocky-reef fish communities owe their greatest diversity in form and colour to the wrasse family (Labridae). More than 90 species have been recorded in temperate Australian waters, ranging in size from the smallest and rarest species, *Pictilabrus brauni* with a maximum size of only 9 cm, to the Western Blue Groper (*Achoerodus gouldii*) which can reach 1.75 m. Several of the larger wrasses are commonly referred to as 'groper' because of their size, even though they are not related to true gropers (family Serranidae). The wrasses that are most typical of temperate waters belong to the genus *Pseudolabrus* and five other closely related groups whose distributions are centred in Australasia. An example is the Crimson-banded Wrasse (*Notolabrus gymnogenis*), one of the most abundant rocky-reef fishes in New South Wales; the males possess characteristic bright crimson fins and may reach 48 cm.

Temperate waters also harbour a small number of species that belong to essentially tropical genera, such as *Coris* and *Halichoeres.* The brightly coloured Combfish (*Coris picta*) is common on coastal reefs of New South Wales. In addition to these resident species, individuals of tropical species such as the Moon Wrasse (*Thalassoma lunare*) can occur at Rottnest Island and the Solitary Islands when the Leeuwin and East Australian Currents carry larvae south. Although juveniles of these species can be seen over the summer months, they seldom survive their first winter.

Unlike their tropical relatives, many temperate wrasse species are found only in Australia and have relatively small geographic ranges. Only a few species, such as the Senator Wrasse (*Pictilabrus laticlavius*) and the Maori Wrasse (*Ophthalmolepis lineolatus*) can be found right across southern Australia. A small number of eastern species can also be found in New Zealand, such as the Luculentus Wrasse (*Pseudolabrus luculentus*) and the Purple Wrasse (*Notolabrus fucicola*). Perhaps the species with the widest distribution is the Inscribed Wrasse (*Notolabrus inscriptus*), which is common on temperate offshore

◁
Combfish,
a subtropical
species found as
far south as
Cape Howe.
Seal Rocks,
New South Wales.
Rudie Kuiter

◁
Maori Wrasse,
one of the few
wrasses found
throughout
temperate
Australia. Sydney,
New South Wales.
Kelvin Aitken

◁
A male
Purple Wrasse.
Rudie Kuiter

islands from Lord Howe Island to Easter Island.

Within Australia, most species are restricted to either the eastern or western coasts, and in some cases, there are sister species on each side of the continent; for example, the Eastern Blue Groper (*Achoerodus viridis*) and the Western Blue Groper (see Chapter 22). Even within regions there can be distinct changes in the common species. For example, in New South Wales the Crimson-banded Wrasse predominates, but is replaced by the Blue-throated Wrasse (*Notolabrus tetricus*) in Victoria and South Australia, and the Brown-spotted Wrasse (*Notolabrus parilus*) in Western Australia. These distributions give the reefs of different regions characteristic mixtures of wrasse species. Wherever you are, there will be many wrasses and at least one wrasse species that is among the most abundant fishes on rocky reefs. The species with the smallest geographic ranges appear to be those associated with subtropical latitudes (eg, Seven-banded Wrasse, *Thalassoma septemfasciata*, which can be found in Western Australia between Rottnest Island and Coral Bay).

▽ Inscribed Wrasse. Bermagui, New South Wales. *Rudie Kuiter*

WHY WRASSES ARE SUCCESSFUL?

There are many reasons why wrasses have become such a dominant part of temperate-reef fish communities. They share with many other reef fishes a highly successful life history involving both a larval and an adult stage. When adults spawn they release their eggs and sperm directly into the water, usually a few metres above the reef. A fertilised egg develops quickly into a plankton-feeding larva that is capable of moving great distances. This life history pattern is successful because it allows adults to cast their offspring over a broad area, ensuring some will find suitable habitat. If a larva just passively drifted in the water currents it is likely that it would be carried a long way from its home reef. However, fish larvae are active swimmers although it is not known whether they use this swimming ability to migrate or just maintain their position in currents that move them around. Wrasse larvae are in the plankton between 20 and 50 days, after which they seek out and settle back into suitable reef habitat. At this stage they may only be 1 cm in length and undergo dramatic morphological transformations. Juveniles look very similar to the adult fish, although in some groups there is a distinctive juvenile colour phase (eg, the Half and Half Wrasse, *Hemigymnus melapterus*). Adult females produce millions of eggs and although mortality through the larval stage is extremely high, a small number usually survive to replenish reef populations throughout their range.

Another reason for the success of the wrasse family must surely be their flexibility in habitat use and diet. They are known to occupy all reef-associated habitats, including kelp forests (eg, the Senator Wrasse), red and green algal-turf areas (eg, Rosy Wrasse, *Pseudolabrus psittaculus*, in Victoria), seagrass beds (eg, Brownfield's Wrasse, *Halichoeres brownfieldi*, in Western Australia), sea urchin Barrens Habitat in New South Wales (eg, Crimson-banded Wrasse) and even sandy habitats adjacent to reefs (eg, Maori Wrasse). Individual species are not necessarily confined to any one of these habitats. Juveniles of the Purple Wrasse for example settle out into the fronds of large brown algae in shallow water but as they grow they move into deeper water and become associated with open areas of reef. Larger species may be capable of moving between habitats on a much broader scale. An example is the Eastern Blue Groper, which is known to recruit as juveniles into seagrass beds within shallow estuaries and move out to exposed rocky reefs where they spend their adult life (see Chapter 22).

All temperate wrasses are carnivores, their flexible and powerful jaws mean that they are capable of consuming almost every kind of

△
Blue-throated Wrasses, are important predator on temperate rocky reefs. Kangaroo Island, South Australia. *Rudie Kuiter*

▽
A Brown-spotted Wrasse, Rottnest Island, Western Australia. *Barry Hutchins*

△ A Crimson-banded Wrasse eating a Blacklip Abalone. Montague Island, New South Wales. *Rudie Kuiter*

invertebrate associated with rocky reefs, from minute copepods to lobsters. Large species can crush the shells of molluscs, including turban shells and abalone and mussels (bivalves), which affords them access to an abundant and energy-rich supply of food. Adults of most species can have quite specialised diets. For example, large Purple Wrasse specialise on different mollusc species at different locations and variation in diet can be determined by individual preferences. Despite their apparent specialisation, wrasses are also opportunists and consume almost any animal food when it is available.

All juvenile wrasses, regardless of their ultimate diets, consume small, relatively soft-bodied crustaceans, mainly copepods and amphipods. These small crustaceans are probably the most abundant animal food source on rocky reefs, living in the fronds of kelp plants or sheltering in the carpets of algae. These animals increase in abundance over summer at the same time that most juvenile wrasses settle out of the plankton into the reef environment. Amphipods represent such a rich source of food that many herbivorous fishes actually feed on them when they first take up residence on the reef.

A unique specialised feeding mode called 'cleaning behaviour' has arisen in the wrasse family, although it is more prevalent in tropical reef communities. Some wrasses feed entirely on skin parasites living on other fishes. Individuals of the Cleaner Wrasse (*Labroides dimidiatus*) spend their time picking over the bodies of larger fishes that visit their 'cleaning stations'. They are known to swim between the gill clefts and the mouth, removing parasites that have infested the gills. Other species, such as the Rosy Wrasse, appear to regularly clean other fishes, although it is a minor part of their diet. While individuals of these two species may clean throughout their lives, others appear to exhibit cleaning behaviour only as juveniles (eg, Comb Fish, Eastern King Wrasse, *Coris sandageri* and the Crimson Cleaner Wrasse, *Suezichthys aylingi*. Cleaning behaviour is often associated with species that have a dark line running along the body, which appears to advertise their services. In the Eastern King Wrasse and the Crimson Cleaner Wrasse individuals appear to lose the dark stripe as they grow into adults and adopt a different mode of feeding.

GROWTH AND AGE

Wrasses in southern Australia generally grow considerably faster than their tropical relatives. Even when food is in short supply, rather than starving to death, individuals simply reduce their growth rate. When food is abundant they appear to grow faster and put more energy into the production of eggs. Length of life is also extremely variable, with small species such as the Blue-throated Wrasse living only seven years in South Australia and to 11 years in Tasmania. The Purple Wrasse lives to 17 years in Tasmania, but has been aged at up to 25 years in southern New Zealand. Thus they appear to have adopted a life history capable of responding to a wide range of environmental conditions, with a tendency to grow slower and live longer in cooler waters.

REPRODUCTION

Wrasses have perfected one of the most bizarre patterns of reproduction in the animal kingdom. Individuals in most species begin life as females and change sex to become males later in life. Associated with this sex change is a bewildering array of colour changes and behaviours. This life history pattern, called 'protogynous hermaphroditism', can be found in other fishes (eg, true gropers), but it is among the wrasses that the greatest diversity of sex change patterns can be observed. The most obvious change associated with the change from female to male in most species is a change in colouration and usually, the male is more colourful than the female. Dramatic colour changes do not come with sex change in all species. For example, males and females of the Black-spotted Wrasse (*Austrolabrus maculatus*) are extremely difficult to distinguish. In most other species, however, there are distinctive male and female body markings. Sex change in some of these 'sexually dimorphic' species (eg, the Western King Wrasse, *Coris auricularis*) is always associated with the external change in colour from female to male.

However, this is just the beginning of the diversity in patterns of reproduction. Sex change can become so complex that there is little relationship between the colour phase and the sex of an individual. Although females never exhibit male colouration, the reverse is not true. In some species, a small number of individuals may be born as males, but through their early life adopt female colouration (eg, Brownfield's Wrasse). In others, a small number of females change sex very early in life but retain the female colouration (eg, Brown-spotted Wrasse). In both these species, when fish change from female to male, they change to the normal male colouration. This means that there are two kinds of males — males that look like females and males that look like males. These two kinds of males exhibit different modes of reproduction. Male-coloured males are usually large, aggressive individuals that defend territories from other males. They tend to actively court females and usually spawn with a single female. In a typical 'pair' spawning run, the female and male swim rapidly upwards, with the abdomens pressed together, and release eggs and sperm into the water above the reef. Female-coloured males, however, tend to exploit the behaviour of the larger males. Using their female disguise they can enter spawning areas, approach pair-spawning fish and join in pair-spawning runs. This behaviour is called 'streaking' for obvious reasons. Females never pair-spawn with female-coloured males. However, groups of female-coloured males are known to harass females into spawning, particularly in species that have a large number of small males. Bands of males adopting this spawning mode are known as 'sneakers'.

The related mysteries of why fish change sex from female to male and why this diversity of patterns in sex change has arisen have not been solved. There is no genetic difference between males and females in these fishes, as there is in many other animals. Also, there are very few morphological differences apart from colour between males and females. Male and female gonads are simply paired structures that either produce eggs or sperm and appear to be able to change from producing one type of gamete to the other with relative ease. However, this does not explain why they change sex. The answer appears to relate to the mating system. In most species a small number of large males dominate the rest and spawn with many different females. A large male has a higher reproductive success than a female of the same size because he can mate with many females, and small males do not get to breed at all except via the subversive tactics described above. A small female has a higher reproductive success than a similar-sized male. You can see then that over a lifetime, if an individual can be female when small and change into a male when it is capable of breeding with many females, it will ultimately produce more offspring than individuals that remain as either females or males. In the course of evolution, this asymmetry in the reproductive potential of males and females has meant that sex change has

△
Combfish cleaning White Ears at a 'cleaning station'. Montague Island, New South Wales. *Rudie Kuiter*

replaced the pattern of reproduction we might consider normal. Although sex change from male to female is also known in some fishes, it occurs in a much more restricted set of circumstances.

But why then do some species have small, female-coloured males? The answer appears to be because in some environments, large males cannot completely exclude all small males from spawning. While it is always an advantage to be male rather than female when large, some individuals can achieve equivalent reproductive success to females when they are small,by adopting the various sneaker–streaker behaviours that I have described. In some species, such as the Purple Wrasse, all fish are born as females, but most sex change occurs before females reach sexual maturity. These species have essentially returned to a 'normal' system in which individuals spend all of their reproductive life as either males or females (a situation referred to as 'secondary gonochorism'). In Purple Wrasses, females may spawn with 20 or more males who aggregate at high-current areas on the reef.

CONCLUDING REMARKS

So few temperate wrasse species have been studied in detail that much of the diversity in ecology and life history remains to be described. The role played by wrasses as consumers of invertebrates on rocky reefs is likely to be extremely important. An increase in the abundance and average sizes of wrasses inside Marine Protected Areas is testimony to the impact that fishing is having on this group. The indirect effect of the exploitation of wrasses on invertebrate communities has yet to be determined. Populations of recreationally and commercially important species, such as the large and long-lived blue gropers and the Baldchin Groper (*Choerodon rubescens*), have been severely depleted in the past. At present, management strategies, such as Marine Protected Areas, bans on harvesting and minimum size limits, may maintain populations at current levels. However, these unique fishes should be closely monitored so that we can respond to future pressures and ensure their conservation in the long term.

▽ Male Western King Wrasse are common on reefs in Western Australia. Rottnest Island. *Barry Hutchins*

Male Eastern Blue Groper.
Merimbula,
New South Wales.
Ken Hoppen

22 Blue groper

Bronwyn Gillanders

Blue gropers are the largest reef fish in temperate Australia. They are docile and diver-friendly fish that are common on many reefs in southern Australia. Their inquisitive nature means that they are popular with divers and large resident fish have become 'local celebrities' on some reefs in urban areas where they are fed by divers. At dive sites such as the Clovelly sea pool in Sydney, the banging of two rocks will often attract Eastern Blue Groper within minutes. Some are so used to human contact that they are able to be touched. Frequently, groups of blue groper can be seen gathered around divers in the hope of food.

EASTERN & WESTERN BLUE GROPER

Blue gropers are wrasses (Family Labridae, see Chapter 21) rather than true gropers (Family Serranidae) as their common name would suggest. They belong to the genus *Achoerodus*. There are two species in the genus: Eastern Blue Groper (*Achoerodus viridis*) and Western Blue Groper (*Achoerodus gouldii*), both of which are confined to southern Australian waters. There are close relatives in the east Pacific and Japan. Although both species were originally described as separate species over 100 years ago, for many years they were viewed as one species. The Eastern Blue Groper is a common member of rocky-reef communities in New South Wales with its distribution extending from Caloundra (Queensland) to Wilsons Promontory (Victoria). The Western Blue Groper is found on the southwest coast between eastern South Australia and the Houtman Abrolhos Islands (Western Australia). Although occasionally it may be seen as far as Port Phillip Bay (Victoria).

As with other wrasses, the juvenile, female and male forms differ in colour in both the eastern and western species. Small juveniles of the Eastern Blue Groper are green, later changing to brown. Females vary between brown and reddish brown and may have a series of pale blotches along their sides, whereas males are grey to blue. A pattern of blue and orange lines radiating from the eyes can be seen on most sizes of fish. By contrast, the Western Blue Groper do not have the series of lines radiating from their eyes. Females are bright green with no series of pale blotches and males vary between green and blue. Each colour form was at one time thought to be a separate species and this explains why groper has in the past been referred to as the blue groper, the brown groper and the red groper. Both the Eastern and the Western Blue Groper have distinctive large, fleshy lips known scientifically as a broad cowel-like upper lip.

The Eastern and Western Blue Groper also differ in their maximum size. The eastern species grows to approximately 1 m, and up to 18 kg, whereas the Western Blue Groper can reach 1.75 m and 40 kg. Many people believe that the colouration and the maximum length that the two species may attain are the only distinguishing characteristics, however, they can also be separated based on the number of scales along and above the lateral line (the lateral line is a sensory system that detects vibrations in the water). The Eastern Blue Groper has 41–45 lateral line scales with 9–10.5 scales above the lateral line. By comparison, the Western Blue Groper has fewer lateral line scales (33–37) as well as fewer scales above the lateral line (7–7.5).

DISTRIBUTION AND ABUNDANCE

Groper do not school but juveniles and females may form loosely knit aggregations of six or more individuals. The majority of fish are seen moving independently and are most active during daylight hours. Like other wrasses they swim using their pectoral fins in a distinctive skulling motion. Only when they need a sharp burst of speed will they use their tail. The use of pectoral fins for swimming also allows groper to be easily distinguished from the similar-looking luderick or blackfish, as the latter fishes swim using their tail rather than pectoral fins.

Small juvenile Eastern Blue Groper are most commonly found in seagrasses or weed habitats in estuaries or sheltered reefs, but they can occur in any sheltered habitat with some sort of physical structure. Peak recruitment to seagrass habitats occurs in spring (September–October), but small fish (<10 mm) can be found during winter, as early as June. Juveniles in seagrasses grow to a length of approximately 50 mm over three to four months before they are thought to move to deeper beds of seagrass

A male Eastern Blue Groper showing radiating lines around the eye and large fleshy lips. Sydney, New South Wales. *Kelvin Aitken*

▷ Male Western Blue Groper. Western Australia. *Rudie Kuiter*

▽ Female Eastern Blue Groper are green to brown. Bermagui, New South Wales. *Rudie Kuiter*

◁ Female Western Blue Groper are bright green and lack the pale blotches of the female Eastern Blue Groper. Western Australia. *Rudie Kuiter*

or directly to coastal reefs. Patterns of abundance of blue groper on reefs in New South Wales have been documented in estuaries around Sydney. Juvenile blue groper occur in their greatest abundances in shallow areas on the reefs inside estuaries. Adult blue groper, especially large fish, are more common on the exposed coastal reefs than reefs inside estuaries. Adults are also found on offshore reefs occurring from very shallow waters to depths of 40–60 metres. These patterns of abundance should be interpreted with care, however, as differences in mortality or growth of fish between estuarine and coastal reefs could account for the apparent patterns of movement. For example, if fish on the reefs inside estuaries are slower growing than the fish on reefs outside estuaries, then small fish from reefs inside estuaries may be of comparable age to large fish from reefs outside estuaries. When growth of fish was investigated, there was no suggestion that fish from reefs inside estuaries grew more slowly than those from reefs outside estuaries.

REPRODUCTIVE BIOLOGY

As is common with many other wrasses, all blue groper start life as females and change into males later in life. Unlike some other wrasses, however, sex change in blue groper occurs after they have bred as females. Colour change (brown to blue in Eastern Blue Groper) is also thought to occur about the same time as sex change, however, some fish may change colour before changing sex, whereas others may change sex before changing colour. Blue colouration may also be size dependent since blue females are, on average, larger than other females.

Females of the Eastern Blue Groper in the Sydney region first mature at a size of 250 mm (0.3 kg), although the majority do not mature until 295–350 mm (0.6–0.9 kg). Whether fish from other parts of New South Wales mature at a similar size is not known. Sex ratios are strongly biased towards females based on observations of the ratio of the number of brown fish ('females') to the number of blue fish ('males'). Female-biased sex ratios are typical of many species of fish in which individual fish start life as females before changing sex.

All Eastern Blue Groper smaller than 600 mm and about 5 kg are females and those greater than this size are males. When the reproductive tissue of blue groper is examined there is little difficulty in distinguishing ripe females, however it is much more difficult to distinguish the ovaries of unripe females from the testes of males and from fish in the process of changing sex. Histological examination is needed to separate these latter phases. Histological analyses proceed by cutting thin sections of gonad that have been embedded in wax blocks; these thin sections are then stained and viewed under a microscope. When histological sections are examined it is possible to see remnants of the female gonads in males and it is from these remnants that biologists know all male fish are derived from females.

Little is known of the factors that control sex change in blue groper. As with other species of wrasse, sex change appears to be a complex process that is influenced by size and age, as well as social and behavioural factors such as interactions among females or between females and males. Although blue groper appear to change sex at a critical size, removals of males and subsequent monitoring of female fish may show sex change occurs at smaller sizes. There is a lack of information on many aspects of the behavioural ecology of blue groper. Behavioural observations of female fish have shown few interactions with other fish of the same species. It is difficult to know whether Eastern Blue Groper have a social hierarchy among females, as is found in some wrasses. If such a social hierarchy exists, the removal of a large male may result in the largest female changing sex. The addition of females to the population may also result in a female changing sex if some threshold between the number of females and males is exceeded. The Western Blue Groper appear to live in small social groups that comprise one male, two to three females and a few immature fish. There are many questions that still need answering on the behavioural ecology and sex change in blue groper.

The Eastern Blue Groper spawn over a three-month period between July and October. There is also some evidence to suggest that large fish may spawn earlier in the season than small fish. Blue groper are unlikely to spawn all at once, as fish with gonads at different stages of development have been found. Little is known about their reproductive behaviour as courting or spawning have not been observed in the wild, despite many hours of behavioural observations. Most of these observations were made during daylight hours so it is possible, but unlikely, that spawning occurs in darkness. Whether blue groper move to deeper areas of water or offshore for spawning is not known.

FEEDING BIOLOGY

Blue groper consume a wide variety of prey items, but are predominantly carnivorous. Small groper in seagrasses and on rocky reefs feed on small crustaceans. On rocky reefs fish change from a diet of crustaceans to one dominated by mussels and sea urchins. Groper attack urchins by flipping them over to reveal the relatively unprotected area around the urchins mouth. A powerful lunge at the urchin cracks it open, and the accompanying thud can be heard quite a distance underwater. Groper may quickly appear when an urchin is offered by a diver and it becomes readily apparent that they are capable of crushing shells or biting at encrusting worms, oysters and barnacles. Other species of fish, such as planktivores, may feed opportunistically around this predation event (see Chapter 25). The size-specific changes in diet seen in blue groper and other species of fish may be due to a variety of factors. Larger fish are better able to crush prey items; therefore as fish get larger, more organisms with hard shells become available.

The habitat in which blue groper forage also changes throughout their life. When small, Eastern Blue Groper on rocky reefs in New South Wales feed in shallow Fringe Habitat but move into deeper water when larger, and feed in turfing algae and in the Barrens Habitat (see Chapter 2). This pattern also reflects the depth distribution of other species in which small fish are found in shallow water and medium- and large-sized fish are found in medium and deep areas of reef.

The main species observed to interact with blue groper is White Ears (*Parma microlepis*, see Chapter 19). This territorial damselfish aggressively chases the much larger groper away from its territories. There is little dietary overlap between these two species although White Ears are likely to protect their breeding territories from the foraging gropers.

△
Male Eastern Blue Groper eating a Black Sea Urchin. Merimbula, New South Wales. *Paul Baumann*

Many other species of fish, such as Crimson-banded Wrasse (*Notolabrus gymnogenis*), Mado (*Atypichthys strigatus*), goatfish (Family Mullidae) and morwong (Family Cheilodactylidae), have been observed either chasing blue groper or being chased by blue groper, but interactions with other species are not common.

AGE AND GROWTH

Eastern Blue Groper have been aged using their ear bones or otoliths. Ageing of fish is similar to looking at a cross section through the trunk of a tree and counting the rings that are visible. Ear bones of blue groper are typically quite small compared to other species of fish, but thin cross sections can be taken and the rings or growth zones counted. Using this methodology, Eastern Blue Groper as old as 35 years have been found. Scales have been used to age Western Blue Groper where a fish of 50 years was found. As expected, the old fish are very large males.

Analyses of size and age data for Eastern Blue Groper shows that an average fish in its second year of life will be 230 mm long (0.26 kg). By the time blue groper are 10 years of age they will be 480 mm long (2.4 kg), at 20 years of age they will be 620 mm long (5.3 kg) and at 30 years they will be 725 mm long (8.4 kg). In South Australia, Western Blue Groper measuring 400 mm in length have been shown to be approximately 8 years old. At 25 years they are 800 mm in length and may be as old as 50 years by the time they reach their maximum size (1420 mm). Hence, very large fish have the potential to be extremely old and may have spent many years on the same reef. The loss of such fish, whether by line fishing or illegal spearfishing, is a highly emotive issue.

PROTECTED SPECIES

Eastern Blue Groper were first protected from fishing in New South Wales in 1969. Prior to 1969, they were targeted by fishers who started taking them in such large numbers that there was some concern for the survival of the species. In the 1950s, for example, 30 per cent of all fish taken in spearfishing competitions in the Sydney region were Eastern Blue Groper. Between 1969 and mid-1974 there was a total prohibition on the capture of Eastern Blue Groper by both amateur and commercial fishers. However, by mid-1974, this prohibition had lapsed, but additional regulations were imposed. The taking of Eastern Blue Groper by spears or similar devices was banned, but a bag limit of two fish/day was allowed for line fishers. In 1975, nine boxes of Eastern Blue Groper were sent to the Sydney Fish Markets highlighting a problem with the legislation that allowed commercial fishers to net groper. Restrictions banning the taking of groper by means of nets were then imposed and a total prohibition on the sale of groper was imposed a few years later. Currently, the taking of Eastern Blue Groper by use of nets of any description and by means of spears, spear guns and similar devices is prohibited.

Various levels of protection exist in other states. Blue groper are not commonly found in Victoria and as a consequence, they are not protected. In Western Australia, where only the Western Blue Groper is found, there is a daily bag limit of one fish which must be larger than 400 mm. The origins of this size limit are unclear as there has been no extensive research on Western Blue Groper in Western Australia. In South Australia the Western Blue Groper is totally protected in a number of areas (eg, Spencer Gulf, Gulf St Vincent), but in other areas there is a daily bag limit of two fish (boat limit of six fish) that must be larger than 600 mm. The origins of this size limit are similarly unclear.

A Horseshoe Leatherjacket at Rottnest Island, Western Australia.
Barry Hutchins

23 Leatherjackets

Barry Hutchins

DIVERSITY AND RELATED FISHES

Australia has more species of leatherjacket than any other country in the world. Of the 97 species found worldwide, 60 inhabit Australian seas and 22 of these are found only in the southern half of the continent. Leatherjackets in temperate Australia inhabit rocky reefs, seagrass beds and sponge gardens at depths ranging from the shallow subtidal to over 250 m. Many leatherjackets spend their early life in shallow seagrass beds, migrating to deeper water coastal rocky reefs as they mature. The young have also been found in floating rafts of *Sargassum* seaweed and amongst the arms of jellyfish. Leatherjackets are mostly colourful fish, with patterns of iridescent blue lines and spots and blotches ranging from yellow, green to white. Some are masters of camouflage, easily blending with the background when danger threatens. They range in size from 9 cm for the Pygmy Leatherjacket (*Brachaluteres jacksonianus*) to about 80 cm for the Chinaman Leatherjacket (*Nelusetta ayraudi*, marketed as Ocean Jacket).

Leatherjackets get their common name from their generally smooth, velvety skin, although some species like the Fan-bellied Leatherjacket (*Monacanthus chinensis*) and the Rough Leatherjacket (*Scobinichthys granulatus*) have coarse skins. All leatherjackets possess small scales covered with minute spines, those of the rough-scaled leatherjackets being more prominent, erect and sharp. Often the skin has a sandpaper-like feel which accounts for the name 'file fish' as they are commonly known in North America.

The main identifying characteristic of leatherjackets is the single spine on the top of the head, usually above the eye. This spine can be raised and lowered when required, and can be locked erect by a second very small spine hidden in tissue just behind the base of the main spine. When down, the spine is sometimes hidden in a groove in the back of the fish,

◁ A male Six-spined Leatherjacket. Montague Island, New South Wales. *Rudie Kuiter*

◁ A Black Reef Leatherjacket on a shallow reef. Sydney, New South Wales. *Barry Hutchins*

especially in members of the genus *Acanthaluteres*. It is often armed with strong barbs, although these become relatively smaller and more worn in old individuals. Unlike the spines of catfish, which are similar in shape, the leatherjacket spine is not venomous. However it is often sharp, and can cause some pain if it penetrates a finger or foot.

Almost all of southern Australia's leatherjackets are endemic to the region, although the Velvet Leatherjacket (*Parika scaber*) is represented by a slightly different form in New Zealand. It is the only species of leatherjacket which is commonly found in New Zealand and differs from its Australian cousin by having slight colour differences and larger scales on the cheeks. There is little doubt that they once had a common ancestor but they no longer interbreed and are slowly evolving into separate species. Other species such as Gunn's Leatherjacket (*Eubalichthys gunnii*) from southeastern Australia has a close relative in the Blue-tailed Leatherjacket (*Eubalichthys cyanoura*) from southwestern Australia. The Variable Leatherjacket (*Meuschenia freycineti*), or Six-spined Leatherjacket as it is known in Victoria, Tasmania and South Australia, is thought to be the same species as its morphology is very similar. The form in Western Australia is slightly different again, but the three are currently considered to represent different geographic variants of the one species, the Six-spined Leatherjacket. The Pygmy Leatherjacket also has two forms, one in New South Wales and one inhabiting Victoria, Tasmania and across to Western Australia. The body shape of the two is slightly different but what is more unusual is their taste. Aquarists who have kept both forms in their tanks report that some predators find the eastern form inedible but not the southern one. Perhaps habitat or dietary differences of the two forms are responsible for this; however, it could also mean that the two forms should be recognised as separate species.

Some southern leatherjackets have relatives in tropical seas. Members of the temperate genus *Meuschenia* have close relatives in the subtropical genus *Cantheschenia*, which includes the Large-scaled Leatherjacket (*Cantheschenia grandisquamis*) from northern New South Wales and southern Queensland. This genus in turn is closely related to the genus *Cantherhines* whose species inhabit tropical coral reefs of the Indian and Pacific Oceans (eg, the Honeycomb Leatherjacket, *Cantherhines pardalis*).

Members of the tropical family Balistidae are commonly known as triggerfishes. They are closely related to leatherjackets but differ by possessing more spines in the dorsal fin (three rather than one). They also have a large bony knob at the rear of the abdomen which is relatively small in leatherjackets (this knob is actually a rudiment of the pelvic fin encased in modified body scales, and is attached to the rear of the pelvis). When a triggerfish swims into a hole to escape a predator, it raises its dorsal spines and lowers the pelvis, thereby jamming itself in the hole. The most primitive leatherjackets — which have more triggerfish characteristics than other species — each have a large robust dorsal spine and a moderate-sized pelvic fin rudiment (eg, members of the tropical genus *Pervagor*). Like all triggerfishes, these primitive leatherjackets also have a flexible joint in the pelvic fin rudiment. A second more advanced group consists of those leatherjackets possessing a moderately robust dorsal spine and a small- to moderate-sized, non-flexible pelvic fin rudiment (eg, the genera *Meuschenia* and *Cantherhines*). The final group consists of the most advanced leatherjackets. These generally have weak dorsal spines and possess either a very small pelvic fin rudiment or completely lack the rudiment. The majority of southern Australia's leatherjackets belong to the second group. The only southern member of the most advanced group is the Pygmy Leatherjacket.

▽ Scanning electron micrograph of the skin of a young Fan-bellied Leatherjacket showing the rows of spines on the body scales which produce the characteristic coarse feel Scale bar = 1mm *Barry Hutchins*

WHERE ARE THEY FOUND?

Leatherjackets of temperate Australia can be divided into three groups based on their ecology. The first group are the shallow-reef dwellers; members of the genera *Meuschenia* and *Eubalichthys* belong here, although several species, including the Stars and Stripes Leatherjacket (*Meuschenia venusta*), Mosaic Leatherjacket (*Eubalichthys mosaicus*), Black Reef Leatherjacket (*E. bucephalus*) and the Four-spine Leatherjacket (*E. quadrispinis*), occur more often on deeper reefs. The second group inhabits deep reefs and sponge gardens, and includes members of the genera *Thamnaconus*, *Nelusetta* and *Parika*, as well as the four species mentioned above. The final group prefers shallow seagrass beds. The genera *Acanthaluteres*, *Scobinichthys* and *Brachaluteres* favour areas of protected seagrasses in large estuaries or embayments, although some species are also found in adjacent rocky reefs and sponge gardens. Irrespective of where the adults live, the juveniles of many species of leatherjackets are well known for utilising shallow seagrass beds as nursery areas, before moving to deeper reefs as they mature (eg, Six-spined Leatherjacket and Yellow-finned Leatherjacket, *Meuschenia trachylepis*). Some species of leatherjacket appear to favour different habitats around southern Australia. For example, the Black Reef Leatherjacket is often sighted on shallow coastal reefs in southern New South Wales but is rarely found in this habitat from Victoria to Western Australia where it prefers deep reefs and sponge gardens to 250 m.

DIFFERENCES BETWEEN MALES AND FEMALES

In most species of leatherjacket, there are large differences in appearance between males and females. Normally, the female has the shape and colour of the juvenile and is generally deeper bodied with rounded fin shapes while the male is more elongate with concave fin profiles (sometimes the male also possesses an elongate, filamentous ray in either the second dorsal or caudal fin). In the Mosaic Leatherjacket for instance, the female is deep bodied like the juvenile, but the male is much more slender. This change in body shape of the male occurs as the fish matures. Some of the bony material of its skeleton is reabsorbed, the bones becoming thinner, especially in the vicinity of the backbone. This decalcification allows the second dorsal and anal fins to move towards each other, pushing against and deforming the underlying bones of the backbone. The male ends up being more torpedo shaped than the dinner-plate shape of the juvenile and female. In addition, the dorsal and anal fin rays (the underlying framework of a fin) of the male become more elongate. The combination of a more slender body and longer fins improves its swimming performance which is probably important in courtship behaviour.

In contrast to these general patterns, male and female forms of the Chinaman Leatherjacket are both slender in shape with elongated

fin rays, but large males often possess slightly deeper bodies than females. Male individuals harvested from deep water during the mating season also possess another unusual feature — they are reddish in colour, whereas the female remains the normal sandy colour. The red colour is due to burst blood vessels in the skin but it is not known if this occurs before or after capture. Possibly, it is another change that is important for mating.

REPRODUCTION

Unlike some fish families, such as the wrasses (Family Labridae) and rock cods (Family Serranidae), there is no evidence that sex reversal occurs in leatherjackets. Even though young male leatherjackets are shaped like the female, their morphology can change quickly once they begin to mature. Furthermore, the sex ratio is roughly one to one for males and females which suggests that there is no sex reversal (in wrasses, for instance, females greatly outnumber males).

Leatherjackets generally produce large numbers of eggs which are fertilised externally by the male. Small species such as the Pygmy Leatherjacket produce only a small number of relatively large eggs. Most southern species start mating during the late winter and early spring period, continuing into summer, although a few species spawn in autumn. The eggs apparently drift in the plankton until they hatch. The small hatchlings of some species actively search out floating objects, while others remain planktonic until settling in their preferred habitat (see page 199).

Some leatherjackets pair up with the same partner for life. The best example of this is the Yellow-striped Leatherjacket (*Meuschenia flavolineata*). Divers often sight one partner first, but the other one is never too far away. Other species form small harem groups in which the males escort a group of females and juveniles. The three members of the genus *Acanthaluteres* are good examples of this. Juveniles of the Chinaman Leatherjacket form large schools when

△
A small mature female Mosaic Leatherjacket showing the dinner-plate body shape. Busselton, Western Australia.
Barry Hutchins

inhabiting embayments such as Cockburn Sound near Perth, Sydney Harbour, and Spencer Gulf. After reaching a certain size (about 180 mm) they migrate into deep water off the coast but still remain loosely associated in small to moderate schools. Other species like the Six-spined Leatherjacket come together in large schools for mating purposes, spending the rest of the year either in small aggregations or as individuals.

The male leatherjacket of some species has large curved spines on the base of the tail. These are used to hit and slash the sides of other leatherjackets, especially during the mating season. Displaying aggression, two males will approach each other head on, quickly raising and lowering their dorsal spines. During this display, they also shake the tail up and down. Each fish tries to angle its body so as to bring the spines at the base of the tail close to the body of their opponent. Upon achieving this manoeuvre, the tail is swung rapidly sideways so that the spines contact the opponent's body, often ripping a large gash in the flesh. After a life of fighting, these spines usually become well worn. Instead of spines, some species have a prominent patch of strong bristles (eg, the Toothbrush Leatherjacket, *Acanthaluteres vittiger*) or an indistinct patch of fine, somewhat elongate bristles, such as is found on the Pygmy Leatherjacket.

ASSOCIATIONS AND EARLY-LIFE HISTORY

Many leatherjackets lead interesting lives, but probably the species with the most fascinating life history is the Mosaic Leatherjacket which seeks out certain species of jellyfish as soon as it is born. If it does not find one, it will quickly die. Those that are successful spend the next few months hiding in amongst the tentacles and under the bell of jellyfish. These small leatherjackets feed on larval crustaceans that make up the diet of the jellyfish. The planktonic crustaceans are stunned by the stinging cells in the jellyfish's tentacles. While this food source is being moved towards the mouth of the jellyfish, it is stolen by the small leatherjackets. As the leatherjackets get bigger, they start to feed on parts of the jellyfish, including the stinging cells which seemingly do not affect the fish (apparently the leatherjacket becomes immune to the sting by a process of acclimatisation). Eventually the leatherjacket will be too big to shelter under the jellyfish and will start to search for a new home. Large juveniles of this leatherjacket are often found around jetty piles and other structures in coastal bays and embayments. After maturation, they move to deeper water seeking rocky reefs (most adults are found at depths between 30 m and 150 m). The closely related Blue-spotted Leatherjacket (*Eubalichthys caeruleoguttatus*) also has a similar life history, but this species is found only in subtropical to tropical waters of Western Australia.

A similar pelagic existence has been recorded in the Toothbrush Leatherjacket (*Acanthaluteres vittiger*). Juveniles of this species seek out floating rafts of *Sargassum* seaweed after hatching and remain with this algae until it floats over a suitable habitat in which to settle. This is usually a shallow seagrass bed in a large estuary or coastal embayment. Here it spends the next six months or so before moving into deeper water (to 40 m) seeking rocky reefs and sponge gardens. The closely related Bridled Leatherjacket (*Acanthaluteres spilomelanurus*) may also share seagrass beds with the Toothbrush Leatherjacket but is never found in floating rafts of seaweed. Instead it spends its first weeks of life drifting in open water as plankton until it finds a seagrass bed where it remains for the rest of its life.

FOOD AND FEEDING

Leatherjackets feed during the day on a broad range of organisms; some species are carnivorous, others mostly herbivorous. The number and type of teeth provide a good indication of their diet. The Chinaman Leatherjacket, for example, is a flesh eater, and is well known for its ability to bite through the shanks of fish hooks and even the finger bones of careless anglers. Other smaller leatherjackets attack small crabs and shrimps but they can also delicately pick off epiphytes growing on seagrasses. These mostly carnivorous species have pointed teeth, whereas those that eat more algae, like the Bridled Leatherjacket, have teeth with truncated cutting edges. Species that scrape plants from the reef's surface usually start out with pointed teeth but these become more rounded as they become worn down with use. The leatherjacket with the most unusual teeth is Paxman's Leatherjacket (*Colurdontis paxmani*) from Western Australia. The truncated teeth in this species are used to eat algae and seagrasses, but there are also a pair of upright tusks on the inner surfaces of the central pair of teeth in the lower jaw. The exact use of these tusks is not known but possibly they are used to strip material from seagrasses.

◁
A male Yellow-striped Leatherjacket. Recherche Archipelago, Western Australia. *Barry Hutchins*

◁
The Net-patterned Jellyfish (*Pseudorhiza haeckeli*) with juvenile Mosaic Leatherjackets around its tentacles. Rottnest Island, Western Australia. *Barry Hutchins*

Despite these specialities, most leatherjackets have a varied diet that includes crustaceans like amphipods and shrimps, polychaete worms, bryozoans, ascidians, hydroids, epiphytes, various algae and seagrasses. Some species also scavenge on the bottom, eating dead fish and shellfish. Seagrass dominates the diet of some leatherjackets but this may only be consumed incidentally as they feed on the epiphytes that grow on the seagrasses. Whether this is so remains contentious because, although the seagrass passes through the fish's gut without being digested in some species, others appear to benefit directly from its consumption.

GROWTH AND AGE

The only well-documented study of the life history of a leatherjacket concerns the South Australian population of the Chinaman Leatherjacket (also known commercially as the Ocean Jacket). This species is short lived and grows quickly reaching a size of 180 mm in the first year. It grows to its maximum size of approximately 800 mm in about 9 years. Mating occurs in offshore waters at depths between 85 m and 200 m in April and May. Females produce about 1 million eggs each year. Juveniles school in shallow bays along the coast and in large embayments, like Spencer Gulf, after which they migrate into deeper offshore waters. Onset of maturity occurs between the ages of two and four years. Males tend to have a higher natural mortality than females, being greatly outnumbered by the latter at five years of age and over. Large individuals are generally females which tend to live two years longer than males.

Leatherjackets belong to a select group of fish including Snapper (*Pagrus auratus*) which develop bony swellings (hyperostosis) on their skull and backbone. These swellings usually form as the fish starts to mature and may get quite large and bulbous, affecting the shape of the fish. In Snapper, this condition results in the large characteristic hump on the head above the eyes, whereas in the Chinaman Leatherjacket, parts of the vertebrae develop enlarged, rounded lumps. These swellings are thought to be sites for calcium storage but their exact function is not known. What is unusual is that some species have these lumps but a closely related, almost identical species may not. For example, the Red Bream of Japan (*Pagrus major*), which is a close relative of our Snapper, has no bony swellings on its skull and vertebrae.

FISHERIES

Many leatherjackets are edible and support commercial fisheries, most notably in New South Wales, although South Australia and Western Australia also harvest moderate numbers. The most important commercial species is the Chinaman Leatherjacket. It used to be very common off the eastern seaboard but overfishing depleted the populations so now most come from South Australia. It supports a trap fishery operating in that state's western waters. Other important species include the tropical Fan-bellied Leatherjacket and the Six-spined Leatherjacket, both of which are common in Sydney Harbour. The Yellow-finned Leatherjacket is another popular coastal species in New South Wales. The Velvet Leatherjacket is taken from deeper waters in southeastern Australia. In South Australia and Western Australia, the Spiny-tailed Leatherjacket is popular fare. Other edible species include the Horseshoe Leatherjacket (*Meuschenia hippocrepis*) and Yellow-striped Leatherjacket. Most leatherjackets are trapped, but netting and trawling are also effective.

CONCLUDING REMARKS

The leatherjacket is one of the most characteristic fishes of temperate Australia's nearshore reefs and seagrass beds. The 22 species provide a truly Australian flavour to the southern coastline; it would indeed be unusual to find an area without at least one species of leatherjacket present. Many of these species are wide ranging, occurring from the east coast around the southern coastline to Western Australia; only a few are restricted to either the east, south or west coasts. The East Australian and the Leeuwin Currents also bring the young of many tropical and subtropical leatherjackets to temperate areas which adds another dimension to leatherjacket diversity. Surprisingly, few Australian leatherjackets have been able to cross the Tasman Sea and become established in New Zealand waters. Only the Velvet Leatherjacket is common in New Zealand, but the Chinaman Leatherjacket and Morse-code Leatherjacket (*Thamnaconus analis*) from Lord Howe Island have been found there on rare occasions. In South Africa, only a few temperate species inhabit its waters, but none occur in South America. Japan has a diverse leatherjacket fauna but most of these are tropical species. The few Atlantic Ocean species are all tropical. Without doubt, the leatherjackets of the southern coastline are uniquely Australian and the most diverse temperate leatherjacket fauna in the world.

Silver Drummer are common over reefs at Rottnest Island, Western Australia. *Sue Morrison*

24 Herbivorous fishes

Geoff Jones

Only about 20 per cent of fishes on temperate rocky reefs are strictly herbivorous, consuming and assimilating only plant material, although many other species include algae in their diet. Herbivory appears to have evolved a number of times in a number of different families of fishes and is often restricted to only a few species within any particular family. For example, of the 10 species in the Family Odacidae in Australian waters, only one species, the Herring Cale (*Odax cyanomelas*) is a herbivore. In contrast, there are six species of drummers (Family Kyphosidae), all of which appear to consume only algae. Other groups, such as blackfishes (Family Girellidae) and sea carps (Family Aplodactylidae), primarily consume algal material, but may be better described as omnivores. While most damselfishes (Family Pomacentridae) have a planktivorous feeding mode, there are eight species of scalyfins (genus *Parma*) in Australia that are largely, but not exclusively, herbivorous (see Chapter 19). Other families, such as the leatherjackets (Family Monacanthidae), include many species with varied diets that also include algae.

While a diverse array of herbivorous fishes can be found in the tropics, including parrotfishes, surgeonfishes, rabbitfishes, angelfishes and blennies, only a few algal-eating representatives of these groups have colonised temperate waters. Surgeonfish of the genus *Prionurus* (sawtail surgeonfish) are common in subtropical areas of Queensland, northern New South Wales and offshore islands. Some other species, including the Convict Surgeonfish (*Acanthurus triostegus*) and Dusky Surgeonfish (*A. nigrofuscus*), are only found in temperate waters as juveniles. The Blue-barred Orange Parrotfish (*Scarus ghobban*) is one of the few tropical parrotfishes whose range extends into temperate waters, both on the New South Wales and Western Australian coasts. Thus, the evolution and importance of the herbivorous feeding mode in temperate waters is an enigma. Why

◁
A male Herring Cale, a common herbivorous fish found around southern Australia. It feeds only on large brown algae. New South Wales. *Rudie Kuiter*

◁
Rock Cale, a common herbivorous fish on exposed reefs of New South Wales. Merimbula. *Ken Hoppen.*

◁
Zebra Fish are easily identified by their barred colouration. *Rudie Kuiter*

is herbivory so much more important in the tropics? Why have many tropical species not managed to colonise temperate waters? And why does herbivory in temperate species have such an unusual distribution among families? We are only just beginning to grapple with these questions.

WHO AND WHERE ARE THEY?

Most of the herbivorous species in temperate waters come from strictly temperate families. The exceptions are the damselfishes and the drummers which are found throughout tropical latitudes. Herbivory in damselfishes is much more developed on coral reefs where this family of fishes has greatly diversified. One tropical species, the Western Gregory (*Stegastes obpreptus*), is common as far south as Rottnest Island, but is only associated with coral-rich habitats. Scalyfins are the only strictly temperate branch of this family. Low-finned Drummer (*Kyphosus vaigiensis*), the most abundant drummer on the Great Barrier Reef, can also be found as far south as Port Hacking in New South Wales. All other herbivorous species have a primarily temperate distribution.

While some herbivorous fish species are broadly distributed across southern Australia, including Herring Cale and Zebra Fish (*Girella zebra*), most have more restricted distributions. The Silver Drummer (*Kyphosus sydneyanus*) has an unusual distribution in that it is common in New South Wales, South Australia and southern Western Australia, but rare in Victoria and Tasmania. It is replaced by the Western Buffalo Bream (*Kyphosus cornelii*) along the West Australian coast between Cape Leeuwin and Coral Bay. As for most reef fishes, there tend to be eastern and western representatives of most herbivorous groups. For example, the Southern Sea Carp (*Aplodactylus arctidens*) can be found in Tasmania and Victoria, and as far as Kangaroo Island in South Australia. The Western Sea Carp (*Aplodactylus westralis.*) extends from Yorke Peninsula in South Australia around to Rottnest Island in Western Australia. A third sea carp species, the Rock Cale (*Crinodus lophodon*) replaces Southern Sea Carp in New South Wales.

Scalyfins are represented by the Big-scaled Parma (*Parma oligolepis*) and the Banded Parma (*P. polylepis*) in southern Queensland and New South Wales, White Ears (*P. microlepis*) and Girdled Scalyfins (*P. unifasciata*) in southern New South Wales, the Victorian Scalyfin (*P. victoriae*) is found along southern Australia and there are three species in Western Australia: Bicolor Scalyfin (*P. bicolor*), McCulloch's Scalyfin (*P. mccullochi*) and Western Scalyfin (*P. occidentalis*). There are two eastern representatives of the genus *Girella* found in the east: the Luderick (*G. tricuspidata*), which prefers sheltered reef conditions and the Eastern Rock Blackfish (*G. elevata*), which prefers wave-exposed reefs. They are replaced by the Western Rock Blackfish (*G. tephraeops*) in South and Western Australia. The discrete distributions of species across southern Australia appear to reflect discontinuities in the coastal current systems that act as a barrier to the transport of larval fishes from one side of Australia to the other.

▽ Rainbow Fish are abundant in kelp forests but unlike the other Australian *Odax* species, they are not a herbivore. Montague Island, New South Wales. *Rudie Kuiter*

HOW MANY ARE THERE?

Although there are few species of algal-feeding fishes, individual species can be very abundant in some locations. In Western Australia, for example, Western Buffalo Bream can account for a large proportion of the biomass of fishes in reef habitats. The local distribution and patterns of abundance of herbivorous species appears to be related to their preferred algal habitats. The number of Herring Cale in an area is correlated with the amount of Common Kelp (*Ecklonia radiata*). In contrast, the abundance of the Victorian Scalyfin increases where there is greater cover of smaller turfing algae and in New South Wales, the local distribution of White Ears is closely related to that of urchin-grazed Barrens Habitat.

Herbivorous fishes differ enormously in their behaviour and patterns of movement. At one extreme are some damselfishes which defend territories from members of the same species and other herbivorous species. In the Victorian Scalyfin, for example, both males and females defend territories that range in size from 3–30 m^2. Scalyfins are highly territorial towards other herbivorous species and where there are large aggregations of scalyfin territories, other herbivores can effectively be excluded. The reasons for these territories is explained in Chapter 19. In contrast, Western Buffalo Bream can be territorial but are also found in large schools. In shallow water, individuals of both sexes can defend polygon-shaped territories of approximately 12 m^2. These territories contain turfing red algae but are bordered by large brown macro-algae, creating a matrix of polygons. In deeper water they appear to move along the edge of the reef in large schools.

WHAT DO THEY EAT?

Many fishes may be primarily herbivorous as adults, but life does not always start that way. It is generally thought that this is so because of the greater need young fish have for protein (as with humans). Some species, such as the Western Buffalo Bream only develop into herbivores once they reach 30–40 cm. At this stage, they develop an elongated gut with associated bacteria, which appear to be necessary to digest the cellulose contained in algae. Some species are not strict vegetarians even as adults. In the Eastern Rock Blackfish, for example, only 70 per cent of the diet consists of algae with the remainder made up of crustaceans. Whether this mix is determined by the degree to which the plant is fouled is unknown. In New South Wales, both the Southern Sea Carp and Luderick feed mainly on understorey and epiphytic red algae, but can consume animal matter.

Many adult herbivorous fishes appear to have highly specialised algal diets. Girellids consume primarily red and green algae, although in most fish there is a small animal component (~15 per cent). Kyphosids, on the other hand, are strict herbivores, consuming a range of red, green and brown algae and are sometimes quite specialised in their diets. For example, Western Buffalo Bream exclusively eat red algae as adults and Herring Cale eat only brown algae, mainly the Common Kelp. It appears to preferentially consume tissue from the primary lamina, whereas the related New Zealand species, the Butterfish (*Odax pullis*) appears to show some selection for reproductive structures located on the secondary laminae. Strangely, the other *Odax* species found on Australian reefs, the Rainbow Fish (*O. acroptilus*) is a carnivore which specialises in eating the small snails that live on kelp plants.

Other species of herbivores are capable of consuming a range of algal types, although they have strong preferences for certain species. For example, the Victorian Scalyfin consumes over 80 species of red and green algae, but has a strong preference for certain fleshy red algal species such as those in the genera *Champia* and *Rhodoglossum*. Individuals defending larger territories appear to have a greater consumption of these preferred species than those on small territories. However, when circumstances arise they can expand their territories to increase the supply of preferred food types. Similarly, sea carps can have highly flexible diets, but their reproductive success is reduced when they are restricted to less preferred species.

HOW DO THEY INGEST AND DIGEST ALGAE?

Differences in diet among herbivorous fishes may in part be explained by differences in their teeth and jaw structure. Girellids and kyphosids, for example, have many rows of closely set teeth that they use for scraping algae from the rock surface or browsing on larger algae. These teeth can be rapidly replaced when they wear out. Sea carps have a wide mouth with numerous small pointed teeth with which they graze the short turfing algae from hard rock surfaces. Herring Cale, on the other hand, have teeth that are coalesced into a 'beak' with a sharp even cutting edge. They use this to excise pieces from the fronds of kelp

▷ Victorian Scalyfin removing drift Common Kelp from its territory. Pope's Eye, Victoria. *Geoff Jones*

▷ Female Herring Cale in the process of clearing a patch of Common Kelp. Note the distinctive bite marks on the plant below her. Sydney, New South Wales. *Neil Andrew*

plants. These pieces are then processed by the action of the pharyngeal apparatus. This structure, in the throat, has fine tooth ridges which chop ingested material into smaller fragments.

Herbivorous fishes typically have an elongated intestine that allows them to assimilate plant material. However, in each group a unique method of breaking down the plant cell walls of the algae has evolved. The Luderick has a highly acid stomach and a characteristic set of micro-organisms for digesting plant material. Kyphosids on the other hand, use microbial fermentation in the hindgut to assimilate cellulose. Herring Cale are interesting because they lack the obvious morphological specialisations for herbivory found in the other groups, such as a gizzard or elongated gut. However, they have dense concentrations of prokaryote and eukaryote microbes in their lower intestines, which appear to function in digestion.

WHAT IMPACT DO THEY HAVE ON ALGAL COMMUNITIES?

Although not diverse, the sheer abundance of herbivorous fishes in some places suggests they may have a significant impact on the amount of algae, and perhaps also the species of algae on reefs. On limestone platforms in Western Australia, for example, the Western Buffalo Bream appears to create a mosaic of patches of red turf-algae, which are separated by stands of large brown algae. It is likely that they are weeding out the larger unpalatable species from within their territory to enhance the growth of the red algae upon which they feed. Herbivorous damselfishes may also be improving the supply of their food species in this way. The Victorian Scalyfin is capable of removing large brown algae from their territories, although they seldom consume this material. Also, by actively defending their territories from other herbivores they appear to be able to maintain a good supply of preferred plant species. Experiments have shown that preferred algae are rapidly exploited by roving herbivorous fishes, once the territorial fish are removed.

Perhaps the most dramatic example of the effect of an herbivorous fish on algae occurs on the New South Wales coast, where at certain times of the year, Herring Cale completely destroy large patches of kelp. We know when, where and how they do, but why they do so still remains a mystery. Each year between August and October, they appear to return to the sites and clear the same patches of Common Kelp by preferentially feeding on the growing region at the base of the primary frond. This means that the plant is destroyed rather than being completely consumed. Patches cleared by Herring Cale are easily identified because for a short period they persist like a forest of tree trunks with no branches or leaves. However, the stipes eventually die and rot away, creating a large bare patch within the kelp forest. Not long afterwards, new kelp plants re-establish and grow to form a new kelp canopy, which is cleared again at the same time the following year.

Why do Herring Cale keep coming back year after year to clear the same patches? One explanation may be that they have a strong preference for one-year-old kelp plants. Once a patch is established, each year they will find a stand of plants aged one year at exactly the same place. This hypothesis was tested by an experiment in which researchers cleared large patches of kelp in other places, thereby establishing new stands of algae containing plants aged less than one year. However, this had no effect on Herring Cale. They continued to return to their traditional feeding sites. It appears that the seasonal impact of Herring Cale is due to a change in the behaviour of females during the spawning season. During the spawning season males defend territories and spawn with females within the territories. Females aggregate at traditional sites near the edges of the territories prior to spawning. They seem to gather in greater numbers near some territories. We do not know whether this is directly because of some attribute of the male or because the location of the territory is better. The clearings appear to be simply a by-product of this dense aggregation of females. Clearance of kelp may not occur at other times of the year because feeding is less concentrated during non-spawning periods.

WHY ARE THERE SO FEW TYPES OF HERBIVOROUS FISH?

Although there are isolated examples of abundant herbivorous species in temperate waters and of species having a major local impact on algal communities, these examples appear to be the exception rather than the rule. The role of herbivorous fishes is much less important than on tropical reefs where grazing fishes abound and have major impacts on algae. This returns us to the question of why so few species have invaded or evolved in temperate waters. One explanation seems apparent. Consuming plant material is known to be less nutritious than consuming a similar volume of animal matter and it is generally assumed that it is harder to 'make a living' feeding in this manner. Tropical species tend to feed on microscopic algae growing in a highly productive environment, which may account for the success of this feeding mode on coral reefs. Grazers with a generalised feeding mode may be able to successfully exploit this food source. The productivity of these kinds of algae appears to decline in temperate waters, which may explain why parrotfishes and other tropical species have not invaded cooler waters. In part, this may be because the rock surfaces where micro-algae grow are shaded by larger kelp species. Thus, to explain why tropical fishes have not invaded temperate waters we might have to first ask the question of why kelp plants have not become widely established in the tropics.

Unlike their tropical counterparts, temperate herbivorous fish generally consume tissue from larger plants and could be considered browsers (taking distinct bites from individual plants) rather than grazers (consuming bits of many individual plants in the same bite). Temperate waters clearly support large standing stocks of macro-algae that would appear to be an abundant source of food. But is it? It appears that many large algal plants contain toxins or other metabolic products that may render them unpalatable to some herbivorous fishes. Brown algae contain polyphenolic compounds (tannins) that are known to act as a chemical defence against browsing fishes. Highly calcareous algae may also be unpalatable to some fish species, although girellids have developed a feeding apparatus and a grinding gizzard that enables them to assimilate these algae.

A third explanation for the depauperate herbivorous fish fauna may lie in the limitations on the feeding apparatus of these fishes. It has been argued that most fish are too small to consume enough algae to meet energetic demands. Studies on sea carps, for example, have shown that only large adult fishes can meet energetic demands by consuming algae alone. Small fishes can only do this by supplementing their diet with animal material or being completely carnivorous as juveniles.

Whatever the reason, understanding why there are so few types of herbivorous fish on temperate rocky reefs is unlikely to be a simple story. Herbivory has evolved a numbers of times under different circumstances, indicating that there may be many different ways to be a herbivore. There is probably no general role of herbivorous fishes in temperate waters because there is no 'typical' feeding mode. The facet of their biology that makes them most interesting is the diverse means by which each species consumes and assimilates plant material. The biology of herbivorous fishes remains one of the most perplexing and poorly understood aspects of temperate rocky-reef ecology.

▽ A patch of Common Kelp that has been recently cleared by female Herring Cale. Note the forest of denuded plants. Sydney, New South Wales. *Neil Andrew*

Ladder-finned Pomfrets schooling over a reef. South West Rocks, New South Wales. *Ken Hoppen*

25 Planktivorous fishes

Tim Glasby and Michael Kingsford

WHAT ARE PLANKTIVORES?

Many of the most colourful and abundant fishes on temperate reefs feed on plankton — tiny organisms that drift in the water. These planktivores often swim gracefully in large schools above rocky reefs as they feed. Some planktivores have a close association with the reef, while others are just temporary visitors from open waters.

Although planktivores generally swim high in the water column, large numbers of some species, such as damselfish in the genus *Chromis*, occasionally swim close to the reef. Some planktivorous fishes shelter in caves and crevices during the day, venturing out only at night to feed. Most planktivores, however, are active during the day when they spend much of their time feeding. Planktivores in turn are an important source of food for larger predators, including humans. Some planktivores rank highly in the total tonnage taken by commercial fisheries. For example, Jack Mackerel (*Trachurus declivis*) are typically in the top three species of fish caught in temperate waters in Australia. These fish are usually caught in open waters rather than close to rocky reefs. Most of the planktivores resident on rocky reefs are not harvested by humans.

The planktivorous fishes come in a range of shapes and sizes. Some of the reef-associated planktivores such as damselfishes are deep bodied and relatively small (less than 20 cm long). Others, such as Blue Sweep (*Scorpis violacea*) and trevally (*Carangoides orthogrammus*), are longer and more slender, and these tend to venture further from shore. The smallest planktivores tend to be found closest to the reef. Typically these species are nocturnal and are found in caves and crevices during the day (eg, Eastern Hulafish, *Trachinops taeniatus*, and bullseyes, genus *Pempheris*).

Some of the most common temperate reef-dwelling planktivores in Australia include Stripeys (*Microcanthus strigatus*), Mado (*Atypichthys*

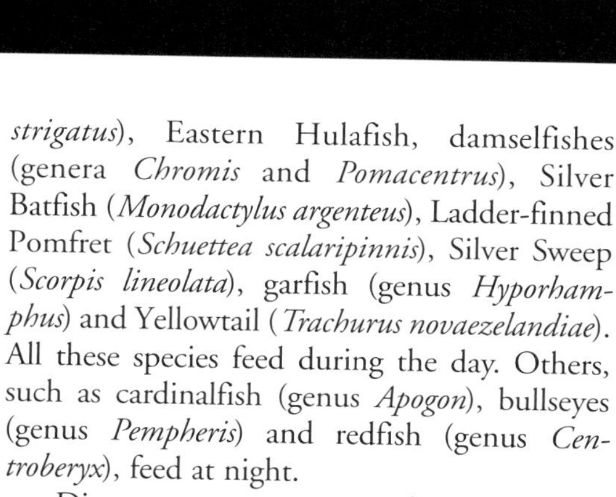

△ A school of Yellowtail above the reef. Montague Island, New South Wales. *Ken Hoppen*

strigatus), Eastern Hulafish, damselfishes (genera *Chromis* and *Pomacentrus*), Silver Batfish (*Monodactylus argenteus*), Ladder-finned Pomfret (*Schuettea scalaripinnis*), Silver Sweep (*Scorpis lineolata*), garfish (genus *Hyporhamphus*) and Yellowtail (*Trachurus novaezelandiae*). All these species feed during the day. Others, such as cardinalfish (genus *Apogon*), bullseyes (genus *Pempheris*) and redfish (genus *Centroberyx*), feed at night.

Divers may see great numbers of vagrant planktivores visiting the reef from open waters. These include trevally and species in the genus *Caranx*, Jack Mackerel, Anchovies (*Engraulis australis*), sprats (*Hyperlophus vittatus*, *Spratelloides robustus* and *Herklotsichthys castelnaui*) and Pilchards (*Sardinops neopilchardus*). More rarely the spectacular Manta Ray (*Manta birostris*), closely related to sharks and with wing spans to 3 m or more, may sweep over inshore reefs. Some species of larger fish often feed on other fish (and are called piscivores), as well as plankton. These species include Red Snapper (*Centroberyx gerrardi*, marketed as Bight Redfish), Longfin Pike (*Dinolestes lewini*), Yellowtail Kingfish (*Seriola lalandi*) and Australian Salmon (genus *Arripis*).

WHAT DO THEY EAT?

Most plankton are less than 2 mm long. These are unable to swim or can swim only weakly against the currents. Common larger planktonic species include some of the jellyfish. Some species of plankton are very ornate — long spines and other structures may provide both buoyancy and some deterrence to predators. Although many species of plankton are planktonic for life, others are the dispersive larvae of larger marine animals such as lobsters, crabs, sea urchins, sea stars and fishes. These larvae may spend days or many months in the plankton before settling near reefs where they spend their adult lives. There are enormous numbers of planktonic organisms in the oceans and they provide abundant food for many animals, particularly planktivorous fishes.

Fishes such as Mado, Silver Sweep, Stripeys and damselfishes often school above rocky reefs during the day. As darkness falls, they move down from the mid-water column closer to the reef. These fish sleep at night and are then most vulnerable to predators. The complexity of the reef, with crevices, caves and rocky overhangs, provides shelter from predators. Conversely, nocturnal planktivores are

▷
Stripeys are common under overhangs in shallow areas of rocky reefs. South West Rocks, New South Wales. *Ken Hoppen*

▷
Red Snapper are nocturnal foragers; they inhabit caves and can be found on deep rocky reefs. Recherche Archipelago, Western Australia. *Barry Hutchins*

▷
Silver Sweep schooling over a reef. New South Wales. *Joachim Ngiam*

◁
Mado feeding around a male Eastern Blue Groper. Seal Rocks, New South Wales. *Kelvin Aitken*

specially adapted to feed in low-light conditions and have characteristically large eyes (eg, Red Snapper) containing large numbers of light-sensitive cells. During the day when they are inactive, nocturnal foragers remain hidden in the reef. The juveniles of these nocturnal foragers, however, do not all feed at night. Young bullseyes feed only during the day because their eyes are not sufficiently well developed to enable them to see during the night.

FEEDING BEHAVIOUR

Many, but not all, planktivores have very small, specialised mouths and acute vision for detecting tiny planktonic organisms. Their narrow heads and close-set eyes enable them to focus on small prey items directly in front of them. Their feeding behaviour consists of quick darting movements with their telescoping jaws which pick individual prey from the water.

Mado are common on rocky reefs in south-eastern Australia and are characteristic of many planktivorous fishes. Although they feed mainly on plankton, they also forage on prey on the reef. Mado often feed around fishes such as Eastern Blue Groper (*Achoerodus viridis*) and Red Morwong (*Cheilodactylus fuscus*) which disturb small reef-dwelling animals while foraging. As they feed, these large fish disperse small animals from the seabed into the water column, where planktivorous fishes can eat them. Planktivores may also surround divers, especially those disturbing the reef. The prey flushed out by large fishes feeding are probably not normally available to planktivores, whose small mouths restrict them to picking prey from the surface of the reef. Planktivores also gain access to much larger prey such as sea urchins and oysters by foraging on leftovers from larger fishes. Opportunistic feeding is common for many planktivores and is an important way for fishes to supplement their diets. This flexibility is important because the supply of plankton is unpredictable.

Some planktivores also behave as cleaner fish. Mado often pick parasites from larger fishes such as Eastern Blue Groper and White Ears (*Parma microlepis*). They are not, however, specialist cleaners like the Cleaner Wrasse (*Labroides dimidiatus*, see Chapter 21) which set up 'cleaning stations'. Larger fishes are sometimes hostile to cleaning by species such as Mado. It is likely that planktivores are not quite as delicate in removing parasites as the specialised cleaner wrasses.

The diet of planktivores varies with the season. For example, during spring in temperate waters fish eggs are abundant in the plankton and many fishes focus on these eggs as food. Diet can also change greatly with distance from the shore. Fishes that feed around offshore islands often eat a greater proportion of krill than those on reefs close to the coast. Many species feed on a variety of organisms, from plankton to algae to small invertebrates in the sediments. Juvenile fish often forage on different prey from those of adults, or the relative proportions of different prey may change as the fish grows. The average size of prey consumed increases as the fish grows. Thus, some species of fish have impacts on a wide range of reef organisms over their lifetimes. The diets of juvenile and adult Mado, for instance, differ substantially. Only large Mado clean other fishes. In some species, feeding patterns may be related to the different shelter requirements of juvenile and adult fishes. Adults generally have less dependence on shelter than juveniles and therefore use different habitats. A combination of differences in size, distribution patterns and social interactions probably cause changes in diet with size. However, fishes of a variety of shapes and sizes can use the same food resources.

THE IMPORTANCE OF PLANKTIVORES ON REEFS

Planktivores have been reported to influence the distribution and abundance of reef-associated animals, both directly and indirectly. The limited information that is available about temperate rocky reef fishes in the southern hemisphere comes mainly from Australia and New Zealand where investigators have described the distribution and feeding behaviours of fishes in addition to associations between fishes and benthic and planktonic assemblages.

Changes in the direction of currents can dramatically alter the abundance and composition of the plankton. Planktivores can generally be assured of finding food in areas where water movements are great. On rocky reefs concentrations of plankton are usually greatest at sites where oceanic waters first arrive. For this reason, many planktivores forage in large aggregations at the 'leading' edge of reefs. Others, especially juveniles, remain in the shallow areas at the back of reefs where there is more protection from predators. Foraging of large numbers of planktivores can reduce local plankton abundance. Large numbers of foraging planktivores have been described as a 'wall of mouths' that can remove most of the zooplankton from the water.

Indirect effects on the organisms on reefs may occur through the deposition of faeces which increase nutrients on rocky reefs. This may be especially important for invertebrates and seaweeds. A rain of faeces descends to the reef from schools of feeding planktivores. Much of this is eaten by other fish. Some faeces are released into the water column at night or directly into cracks and crevices. The latter are particularly important to benthic organisms. By feeding and sleeping in different types of places, fishes may import nutrients to the reef from hundreds of metres away.

Planktivorous fishes are also a food source for other fishes. Planktivores are the most important prey for many carnivorous fishes because of their abundance on reefs. Both juvenile planktivores and adults are eaten by large predatory fishes. Some of those predators are residents of temperate reefs, such as wirrah (genus *Acanthistius*), seaperch (genus *Hypoplectrodes*), Sergeant Bakers (*Aulopus purpurissatus*), rockcod (genus *Scorpaena*), Beardies (*Lotella rhacina*) and moray eels (genus *Gymnothorax*). Other predators are transient and include Yellowtail Kingfish, Longfin Pike and sharks. Trevally are a favourite food of Grey Nurse Sharks (*Carcharias taurus*).

DISTRIBUTION AND ABUNDANCE

Small planktivores often inhabit shallow, complex parts of the reef and do not venture far up into the water column where larger fishes feed. Large differences in types of planktivores may occur between mainland reef and those offshore. For example, One-spot Pullers (*Chromis hypsilepis*) are one of the most common fish at Montague Island in New South Wales, but are scarce on reefs along the adjacent mainland. Silver Batfish are very common in northern New South Wales, but are relatively rare in southern New South Wales and Victoria. Conversely, some species of sweep are abundant in Victoria, but rare in New South Wales. In northern New South Wales there are species of planktivores that are best known in tropical regions, but persist as adults in these temperate areas.

Juveniles of some more tropical planktivores such as Sergeant Majors (*Abudefduf vaigiensis*) and Humbug Dascyllus (*Dascyllus aruanus*) may disperse south as larvae and settle on rocky reefs, but the juveniles often die during the cooler winter months. The appearance of these seemingly exotic species is more common at offshore islands such as the Solitary Islands and Rottnest Island than at equivalent locations on reefs near the mainland.

Some planktivores, such as Jack Mackerel, trevally and Pilchards, are known to move considerable distances over the continental shelf and among reefs on islands and the mainland. Most planktivores, however, appear to be resident on reefs and rarely move to distant reefs (kilometres apart). They may, however, move hundreds of metres within a reef system to find good feeding, spawning and sleeping sites. For example, diurnal planktivores, such as

▽ Yellowtail. Busselton, Western Australia. *Barry Hutchins*

damselfishes, sleep in crevices at night and may migrate hundreds of metres to feeding sites where the supply of plankton is good. When they are in a good area for feeding they may move tens of metres to hundreds of metres depending on the direction of currents, because they prefer to feed on the leading edge of reefs. Juvenile do not make these movements, possibly because they are far more vulnerable to predation. Nocturnal planktivores also move. They often shelter in large aggregations during the day and move to good feeding sites at night. Bullseyes, for example, move to shallow water to feed, but remain hidden in caves in deeper water during the day. In some parts of the world, nocturnal planktivores move hundreds of metres in search of food. Again, the smaller nocturnal species tend not to move too far from their shelters to feed.

REPRODUCTION

Divers are more likely to see spawning by planktivores than by any other type of fish. Almost all of the planktivorous fish in temperate waters around Australia spawn thousands to millions of eggs every spring and these are fertilised in the water column by the males. The eggs are not protected and are very susceptible to predation. Large numbers are lost to planktivores, but enough eggs are produced to ensure that many will survive.

Some fishes (eg, Trevally, Jack Mackerel and Silver Sweep) spawn high in the water column and the fertilised eggs drift away from the reef before the larvae hatch. Others, for instance damselfishes, spawn the eggs directly onto the reef. Most damselfishes normally feed in the water column, but during the spawning season large numbers of them may be found very close to the rocky reef. The males often set up a territory; there they display to females and encourage them to spawn. After mating, the males vigorously guard their clutch of eggs (each egg is about 1 mm long and can be seen by divers). After as few as five days the eggs hatch, generally after dark. Spawning may occur a few times within the reproductive season, with only a week or two in between. The males spend this time in the water column feeding to prepare themselves for the next episode of guarding the nest.

CONCLUSIONS

Planktivores are some of the most conspicuous fishes on temperate rocky reefs. Large aggregations are common, but the location of schools changes greatly according to the direction of currents, supply of food and time of day. With the exception of larger species such as Jack Mackerel and Silver Sweep which take bait, many reef-dwelling planktivores may not be well known because they are generally small and are rarely taken by line or spearfishers. Certainly we have a lot more to learn about planktivorous fishes, but it is clear that they interact in a variety of complex and important ways with other organisms on temperate rocky reefs.

▽ One-spot Pullers are often abundant on rocky reefs. Jervis Bay, New South Wales. *Ken Hoppen*

Australian Fur Seals in typical underwater poses. Wilsons Promontory, Victoria.
Ken Hoppen

26 Seals and sea lions

Peter Shaughnessy

Seals, fur seals and sea lions form a group of amphibious mammals called the pinnipeds or Pinnipedia (meaning fin-footed). They belong to two families, the eared seals (Family Otariidae), which includes the fur seals and sea lions, and the true or earless seals (Family Phocidae). Seals, fur seals and sea lions are amphibious in the sense that they breed, moult and suckle their young on shore (or on pack ice), but feed entirely at sea. Aside from possessing external ears, fur seals and sea lions may be distinguished from true seals in that they use both fore and hind limbs for moving on land, whereas true seals do not use their hind flippers to move when out of water.

The pinnipeds are usually recognised as an homogeneous group which arose from a common ancestor about 14 million years ago. An alternative view is that the two families arose from different ancestors: the eared seals from a bear-like ancestor on the shores of the eastern North Pacific Ocean and the true seals from an otter-like ancestor on the shores of an ancient inland sea (Tethys) in eastern Europe and western Asia.

Three species of eared seal are resident on the coast of southern Australia: the Australian Sea Lion (*Neophoca cinerea*), New Zealand Fur Seal (*Arctocephalus forsteri*) and Australian Fur Seal (*A. pusillus doriferus*). In addition to these resident species, several other species are occasionally seen on the Australian coast; the most frequent are Subantarctic Fur Seals (*A. tropicalis*) and Southern Elephant Seals (*Mirounga leonina*), which breed on the Australian subantarctic islands (Heard and Macquarie), and the Leopard Seal (*Hydrurga leptonyx*), which breeds on pack ice in the Southern Ocean. The latter two species are true seals.

AUSTRALIAN SEA LIONS

The Australian Sea Lion is one of the most attractive seals; adult females and juveniles are ashy grey to pale brown above and creamy

◁ A group of young Australian Fur Seals. One animal clearly shows its external ears (pinnae), and another has a piece of netting around its neck, an 'entanglement'. Kanowna Island, Wilsons Promontory, Victoria. *Ken Hoppen*

◁ Adult male Australian Sea Lion on Neptune Islands, South Australia. *Ken Hoppen*

yellow below, adult males are chocolate brown with a pale mane. Their relatively blunt noses provide a simple way of distinguishing them from the sharper-nosed fur seals. Adult male sea lions reach up to 2.5 m long and weigh up to 300 kg, but females only reach 1.8 m and 100 kg.

The Australian Sea Lion is the only seal species that is found solely in Australia. It breeds on the west and south coasts of Western Australia, and in South Australia. There are 66 breeding colonies on islands and around the coast between the Houtman Abrolhos Islands in Western Australia to The Pages Islands, near Kangaroo Island in South Australia. There are also recent records of a few vagrants on the New South Wales coast, in southern Tasmania and Victoria.

Australian Sea Lions breed on an 18-month cycle and have an extended pupping season of five or more months rather than the more typical patterns for seals of an annual breeding season lasting a month. Pregnancy in Australian Sea Lions is also longer, lasting 14–15 months. Mothers suckle their pups for 18 months until the next pup is born, or even longer if the next pup dies. Other seal species nurse for much shorter times. Breeding seasons of individual colonies of the Australian Sea Lion are not in synchrony, and do not seem to vary in any consistent manner. For other seal species, breeding generally occurs at similar times, in either spring or summer, although the timing may vary with latitude in some species. There are an estimated 11 000 Australian Sea Lions in Australia, 70 per cent of which are in South Australia and the remainder in Western Australia. Most seal colonies are small, with fewer than 100 pups but 40 per cent of the total population is in three colonies in central South Australia, at Dangerous Reef near Port Lincoln, Seal Bay on Kangaroo Island, and The Pages Islands near Kangaroo Island. Of these, The Pages Islands supports the largest population with 400–450 pups produced each pupping season.

Australian Sea Lions feed on a wide variety of prey, including fish, cephalopods, rock lobsters and sea birds. There is little quantitative information on their diet because little of their prey can be identified in faeces. They are known to feed at fishing boats and also to rob bait from rock lobster pots and juveniles occasionally become trapped in the pots and drown. Although Australian Sea Lions are a nuisance to the fishing industry, they are of value to tourism at several sites in South Australia and Western Australia. At Seal Bay on Kangaroo Island, sea lions have become habituated to people and 100 000 people visited the site in 1996. Sea lions are also found at non-breeding sites near Perth and occasionally haul-out (come ashore) in the Perth metropolitan area.

NEW ZEALAND FUR SEALS

Fur seals differ from all other species of seal in having two layers of hair in their pelt: a coarse outer layer and a fine, dense inner layer of fur. The inner fur layer traps air and remains dry when fur seals dive and contributes to their insulation. It was these pelt characteristics that made these seals so eagerly sought after by sealers and that formed the basis of one of the earliest industries in Australia. One of the distinguishing features of this species is their long pointed, narrow snouts, which are especially prominent in adult males, more so than in Australian Fur Seals. Adult males are larger than females, attaining a size of 2.5 m and 180 kg, while females reach 1.5 m and 50 kg.

In Australia, New Zealand Fur Seals breed on islands on the south coasts of Western Australia and South Australia, and at Maatsuyker Island and Macquarie Island. Most (77 per cent) of the population on the Australian coast breeds in central South Australian waters, from Kangaroo Island to southern Eyre Peninsula. They also breed in New Zealand (primarily on the South and Stewart Islands) and its subantarctic islands. The breeding season is in early summer. In South Australia, 90 per

A colony of Australian Fur Seals on Kanowna Island, Wilsons Promontory, Victoria. *Ken Hoppen*

△
A young Australian Sea Lion nursing from an adult female, possibly its mother.
Seal Bay, South Australia.
Ken Hoppen

cent of pups are born in a 30-day period centred around 21 December. Adult females give birth soon after coming ashore in December, mate eight days later and then leave the colony two days later to feed. They lactate for several months, alternating periods at sea feeding with suckling their young ashore.

There are occasional reports of non-breeding animals from the west coast of Western Australia (including the Perth metropolitan area), Victoria, Bass Strait islands, New South Wales (particularly Montague Island) and Queensland (southern part of Fraser Island). Animals on the east coast of Australia may have moved there from New Zealand or from South Australia. The only evidence of trans-Tasman movement is a tagged seal that drowned in a net off Lakes Entrance, Victoria in October 1994; it had been tagged as a pup in a New Zealand colony. New Zealand Fur Seals with coloured flipper tags have been seen on the New South Wales coast at Tathra, Montague Island, Jervis Bay and Sydney. These seals are likely to have come from Kangaroo Island, South Australia, where such tags have been used.

In the early 1990s the population size in Australian waters was estimated at almost 35000 animals. Over the last 10 years, the population size at breeding colonies on Kangaroo Island has been increasing at 10 per cent annually, and increases have also been reported at other colonies. Consequently, the total population size in Australia is now likely to be close to 80 000.

New Zealand Fur Seals take a wide variety of prey, principally fish in summer and cephalopods in winter. Their diet also includes Little Penguins (*Eudyptula minor*) and other seabirds. At Kangaroo Island during the early part of lactation, females dive in shallow waters on the continental shelf to depths of 70–80 m. In late lactation (during winter) dive depths are more variable, between 20 m and 200 m near the shelf edge or just beyond it. They dive at night and rest at the sea surface during the day.

AUSTRALIAN FUR SEALS

The Australian Fur Seal is recognised as a subspecies of the South African Fur Seal (*Arctocephalus pusillus pusillus*). Although they have a paler coat than New Zealand Fur Seals and the snout is less prominent, the two species are difficult for the casual observer to tell apart. Australian Fur Seals are considerably larger than New Zealand Fur Seals, males reaching 2.3 m and 360 kg, while females grow to 1.7 m and 110 kg. Given this difference in size, it is not surprising that Australian Fur Seals are more ungainly on land and usually move one fore flipper at a time and rotate their head and neck as they sway from side to side. In contrast, the more agile New Zealand Fur Seals move forward on both fore flippers together. Vocalisations are also useful for distinguishing between the two species. Calls of the Australian Fur Seal are lower pitched than those of the New Zealand Fur Seal which has a trill associated with its call. Further, molars of Australian Fur Seals have small cusps on the sides next to the adjacent teeth, which are absent in other fur seal species.

Australian Fur Seals only breed on islands in Bass Strait: four in Victoria and six in Tasmania. Their foraging range includes South

Australia, southern Tasmania and New South Wales; several haul-out sites are known in each state. The breeding season is during early summer and pups are born from late October to late December, with most being born in early December. Adult females give birth soon after coming ashore, mate about six days after giving birth, and then leave the colony to feed. They alternate periods at sea feeding with suckling the pup ashore for several months. There is considerable variation in the time of weaning. Pups begin to forage effectively in June or July, supplementing their milk diet. Most are weaned by October, but a small proportion continue to suckle into their second year.

In 1991, the population size for Australian Fur Seals was estimated at between 47 000 and 60 000 animals. Numbers of pups increased in four of the five Tasmanian colonies during three years in the early 1990s but the generality of this increase in other colonies in Australia is unknown. For the largest colony, at Seal Rocks in Westernport, Victoria, pup numbers increased at 2 per cent per annum from 1967 to 1991, and then at 6 per cent per annum to 4200 in the six years to 1997. Despite the recent increases, the overall population size of the Australian Fur Seal is likely to be lower now than it was historically.

The food of Australian Fur Seals is varied but they eat mostly fish in summer and cephalopods in winter although seabirds are also eaten. The most important squid taken in Tasmanian waters is Gould's Squid (*Nototodarus gouldii*) and the most important fish are Redbait (*Emmelichthys nitidus*), leatherjackets (Family Monacanthidae) and Jack Mackerel (*Trachurus declivis.*) Australian Fur Seals are known to dive to at least 200 m. They also feed at fishing boats and are sometimes caught in trawl nets; video cameras have shown them swimming adjacent to trawl nets on the bottom at depths of up to 80 m. The quality of the images was insufficient to show whether or not the seals were swimming within the nets. Entanglement in nets and plastic packaging is a problem in many seal species and seems to be so for Australian Fur Seals, in which almost 2 per cent of the population is reported to be affected.

SUBANTARCTIC FUR SEALS

Adult male Subantarctic Fur Seals have distinctive chocolate brown to black fur on their upper surface contrasting with a yellow chest and face. A dark crest on the top of the head becomes erect when they are annoyed. Adult females have similar colouring but the contrast is not as marked and the females have a short nose which gives them a pug-like appearance. Adult males grow to be 2.0 m long and 160 kg, while females reach 1.4 m and 50 kg.

▽ An adult male New Zealand Fur Seal illustrating the long pointed snout of this species. Kangaroo Island, South Australia. *Liz Poon*

Subantarctic Fur Seals breed in small numbers at Macquarie Island in November and December. The largest colonies are on the subantarctic islands of Gough (South Atlantic Ocean) and Amsterdam (South Indian Ocean). There are many records of these seals ashore in southern Australia, from Western Australia to New South Wales. Most animals of this species at Macquarie Island have been tagged but none of the visitors to the Australian coast have been reported as tagged so Macquarie Island is an unlikely source of these animals. The most likely source of Subantarctic Fur Seals to the Australian coast is Amsterdam Island which is approximately 3600 km from Cape Leeuwin on the Western Australian coast.

The population of Subantarctic Fur Seals at Macquarie Island is small and probably less than 100. Estimates of numbers on the island are confounded because they hybridise there with Antarctic Fur Seals (*Arctocephalus gazella*) which are rarely seen further north. In the summer of 1995–6, 123 fur seal pups were born and of these, 25 were Subantarctic Fur Seals and nine were hybrids. The breeding population at Macquarie Island is increasing at about 12 per cent annually.

At Macquarie Island, they feed almost entirely on pelagic myctophid fish (genera *Electrona* and *Gymnoscopelus*). Recent studies of foraging behaviour at Macquarie Island indicate that these seals forage at night, usually at shallow depths. It will be interesting to see how these seals interact with the fishery that has recently begun near Macquarie Island.

SOUTHERN ELEPHANT SEALS

Southern Elephant Seals are the leviathans of the pinnipeds. Males with their distinctive enlarged proboscis attain a size of 4.2 m and almost 4 t and, although females only reach 2.6 m and 350 kg, they are still larger than most other seals. Despite their large size, Elephant Seals are capable of rapid (but brief) movement on land.

Southern Elephant Seals breed on subantarctic islands including Macquarie and Heard Islands. There are also populations of these seals at other subantarctic islands, including a large population at South Georgia, in the southern Atlantic Ocean. Southern Elephant Seals formerly bred on islands in western Bass Strait, but were eliminated by sealers in the early nineteenth century. Pups are born during

◁
An adult male Subantarctic Fur Seal displaying the yellow chest and face that contrasts with the darker fur on its upper surface. Macquarie Island. *Peter Shaughnessy*

▽
An adult male Southern Elephant Seal showing its enlarged proboscis. Macquarie Island. *Peter Shaughnessy*

September and October, and are suckled for three weeks while their mothers fast. They are then abandoned by their mothers and remain for several weeks ashore before departing to sea.

Studies of Elephant Seals fitted with time–depth–temperature recorders have shown that some of them move far south, and forage close to the Antarctic coast. Elephant Seals visit the Australian coast, particularly Tasmania, where they are frequently reported and several pups have been born. There are also records of them ashore in New South Wales (including two near Sydney) and several in Victoria, South Australia and Western Australia. Many of the Elephant Seals ashore on the Australian coast between December and March are moulting. They fast during the 30–40 day moult, during which hair and skin are shed, and they are reluctant to go to sea.

In the 1980s, there were 86 500 Elephant Seals at Macquarie Island and 13 000 at Heard Island. Population size decreased at these islands in the three decades to the early 1980s. The causes of these declines are unknown and the Macquarie Island population now appears to have stabilised.

Southern Elephant Seals feed mainly on cephalopods and fish. Their major foraging areas are in cold Antarctic waters, along the Antarctic Polar Front and in warmer subantarctic waters north to 50°S. They have not been reported to interact with fisheries on the Australian coast, but one was caught in the trawl net of an Australian vessel fishing in subantarctic waters for Patagonian Toothfish (*Dissostichus eleginoides*).

LEOPARD SEALS

Leopard Seals have a distinctive slender, almost reptilian appearance with a head that seems disproportionately large and a large mouth with many teeth. Their coat is dappled grey and is paler on their bellies than their backs. They move on land by wriggling, and their small fore flippers do not touch the ground. They are large animals, exceeding 3 m in length and reaching 500 kg. Adult females are larger than males, but differences between males and females are not as great as those found in the land-breeding seal species.

Leopard Seals breed, moult and rest on pack ice, and their movements are associated with the seasonal expansion and contraction of ice. They are frequent visitors to Macquarie

Island and to the Australian coast, particularly Tasmania. These animals are usually emaciated. Records of such visitors appear to peak on a cycle of 4–5 years. Their abundance in the Southern Ocean has been estimated at between 222 000 and 440 000 from aerial and shipboard censuses in the pack ice. There is no information on trends in abundance. Leopard Seal pups are born from October to mid-November, and mating occurs during December and early January. Lactation lasts for up to four weeks.

Leopard Seals have a reputation as fearsome predators of warm-blooded animals, such as penguins and other seals, but their main prey is plankton, especially krill. In Antarctic waters, they have been known to chase divers, harass small boats and to attack humans on the pack ice. Their calls may be heard long distances underwater, especially during the breeding season.

REPRODUCTIVE BEHAVIOUR

The fur seals described here are social animals that breed on land and in colonies. Adult males establish territories before females come ashore and aggregate for the breeding season. The males stay ashore fasting for several weeks and defend their territories using growls, stares, bluff and, to a lesser (but more spectacular) extent, by fighting. In most instances, an adult male seal will hold a territory containing several females with which he hopes to mate when they come into oestrus a few days after whelping. Some of the younger males, which are reproductively mature but not yet socially mature, stay on the edge of the colony, or make forays into other territories in the hope of finding a receptive female before the territory holder finds them. Most mating takes place on land, with possibly a small amount occurring in the shallows adjoining the colony.

Several features are associated with this mating system: males are larger than females, males take no part in raising offspring, males are older than females when they enter the reproductive part of the population and are reproductive for fewer years, and females live longer. Curiously, Australian Sea Lions at Seal Bay, Kangaroo Island, display a variation on this theme. Territorial males are sexually monogamous, attending to one female at a time rather than a group of females. They aim to mate with several females over the duration of the breeding season (which is relatively long in this species).

MARINE ENVIRONMENT

The marine environment over much of the range of seals in Australia is characterised by shallow and relatively unproductive waters on the continental shelf (< 200 m). In western South Australia and Western Australia the Leeuwin Current feeds warm, nutrient-poor waters southwards along the west coast of Australia and then eastward along the south coast. This current acts as a barrier to the rich subantarctic waters; the region has been described as being one of the most nutrient-poor marine environments in the world. During winter the prevailing winds along southern Australia are westerly and, as the Leeuwin Current flows most strongly then, the current reaches its eastern extremity. During summer the high pressure weather systems that dominate the south coast of Australia cause consistent southeasterly winds that have the effect of blocking, and in some cases reversing, the flow of the eastward-moving Leeuwin Current. This also facilitates minor upwellings of relatively nutrient-rich, cool water. These influences result in more productive waters in the eastern part of the range of the Australian Sea Lion and the New Zealand Fur Seal.

The main habitat of the other resident species, the Australian Fur Seal, is Bass Strait and, to a lesser extent, the waters of Tasmania and southern New South Wales. In general terms, the waters of Bass Strait are dominated by warm, nutrient-impoverished water derived from the west and north, with seasonal variations. It is only off the east coast of Tasmania and off southern New South Wales that productivity is enhanced by upwellings of nutrient-rich water.

Although fur seals and sea lions are a relatively common sight on reefs in some parts of southern Australia, especially close to breeding colonies and haul-out sites, they are unlikely to play an important role in the dynamics of rocky-reef ecosystems because they mostly forage further offshore.

DIVING

Seals are good divers, particularly Southern Elephant Seals, which can dive to depths of up to 600 m. Furthermore, they dive continuously, with 90 per cent of their time at sea spent diving and average dive duration for individual animals ranging from 16–37 minutes. Seals have several mechanisms that allow them to spend such long periods underwater. Remarkably, seals *exhale* immediately before diving. When they dive, what little air remains in their lungs is pushed into the air passages, where there is little gaseous exchange because there are few blood vessels. Compared with terrestrial mammals, they have an enhanced capacity to store oxygen in their blood via a modified circulatory system. Blood makes up a greater proportion of their body weight, red blood cells are more abundant and the concentration of the protein that combines with oxygen (haemoglobin) is greater. The corresponding protein in muscle, myoglobin, is more abundant than in terrestrial mammals, and gives seal meat its characteristic dark colour.

Seals are also able to slow their heart rate to fewer than 10 beats per minute when underwater. This ability, or bradycardia as its known technically, is slight in terrestrial mammals, but pronounced in marine mammals and is accompanied by a redistribution of blood flow so that only the essential organs (brain and heart) receive oxygenated blood. Large volumes of blood are then stored in sinuses within the abdomen and thorax. Seals are also able to effectively store the products of anaerobic metabolism (lactic acid and carbon dioxide) during long dives, which they then rid themselves of when they breathe again at the surface.

STRANDED SEALS

Seals that come ashore away from colonies and regular haul-out sites are often reported as being 'stranded'. These seals may be resting in unusual places because they are sick, but this is not necessarily the case. The word stranded is not always appropriate for seals because it implies that the seal has not come ashore of its own volition and that it requires assistance. Seals may be ashore for many reasons, and may not require assistance. For seals beyond their normal range, the terms 'vagrant' or 'extralimital' are more appropriate than stranded. Nevertheless, there is no doubt that 'stranded' and 'stranding' will continue in general use for seals, and they are used here for convenience.

If the seal looks healthy or if it is moulting, it is best left alone. This includes not chasing it into the sea; such action is dangerous because seals can be mobile on land and can inflict serious injury by biting. Stranded seals should be reported to the local nature conservation agency or police. People who feel obliged to 'do something', should endeavour to keep other people and dogs at a distance until qualified help arrives.

Seals ashore should not be fed, because of the likelihood of providing inappropriate or contaminated food, and the danger to the feeder. It is illegal to handle or harass a seal without a permit. It can also be dangerous, despite the friendly images conveyed by the media. There is also the danger of diseases that are transmissible to humans, such as tuberculosis, 'seal finger' (a severely painful infection with associated swelling) and 'seal pox' (a skin disease, with skin nodules about 1–2 cm in diameter).

Seals can recover from major flesh wounds without human intervention so unless there is convincing evidence that a seal is in distress, it is best left alone. Signs of distress include obvious emaciation, hyperventilation (possibly caused by a plastic bag caught in an airway), a deep, large wound (possibly caused by a boat propeller or a shark), discharge from the nose and being entangled.

INTERACTION WITH FISHERIES

Interaction between seals and fisheries may be overt or indirect. Seals attend fishing boats and fishing gear, take fish that have been caught, take bait, disperse schools of fish targeted by fishers or drive them beyond the range of nets, and damage equipment. These interactions may have fatal consequences for seals: some are shot and others become entangled in fishing gear and discarded plastic packaging. On the Australian coast, the most obvious interactions with seals are those involving set nets, such as those used to catch Australian salmon (genus *Arripis*) on the south coast of Western Australia and to catch sharks in South Australia, as well as the drop-line fishery in Tasmania; the rock lobster fisheries in southern Australia; and fish aquaculture in Tasmania. In all these fisheries the fishing gear is relatively static, giving seals ample time to find the nets and interfere.

Seals interact with fisheries more indirectly by competing for common prey species. Quantitative information on competition between seals and fishers is required to determine the extent to which seals and humans are competing for the same prey. In New Zealand, for instance, there is little overlap between the prey of New Zealand Fur Seals and fish species taken in commercial fisheries. On the other hand, an important prey item of Australian Fur Seals is Jack Mackerel of a similar size to those taken commercially. When considering ecological interactions between seals and fishers, the complex interactions between predators and prey should not be overlooked. For instance, predators of rock lobsters include octopuses and numerous fish species, which in turn are preyed on by fur seals and sea lions on the Australian coast.

Further reading

The following publications are designed to provide the interested reader with an entry point to the scientific literature. The list is not intended to be exhaustive. Although much of the work done on temperate rocky reefs remains unpublished in theses by postgraduate students, these have not been cited because they are difficult to obtain. The other major bibliographic source is final reports to the Fisheries Research and Development Corporation (FRDC). The FRDC maintains a complete library of the results of FRDC-funded research, and non-technical summaries of the reports are available through their web site (http://www.frdc.com.au). The publications below are listed by relevant chapter. In addition, identification guides and publications of general fisheries or ecological interest are given.

IDENTIFICATION GUIDES

Allen GR (1991). *Damselfishes of the World.* Mergus: Melle, Germany.

Edgar GJ (1997). *Australian Marine Life: the plants and animals of temperate waters.* Reed Books, Sydney.

Fautin DG and Allen GR (1992). *Field guide to anemonefishes and their host sea anemones.* Western Australian Museum, Perth.

Hutchins B and Swainston R (1986). *Sea Fishes of Southern Australia.* Swainston Publishing, Perth.

Kuiter RH (1993). *Coastal Fishes of South-eastern Australia.* Crawford House Press, Bathurst, NSW.

Kuiter RH (1997). *Guide to Sea Fishes of Australia.* New Holland Publishers Pty Ltd, Sydney.

Last PR and Stevens JD (1994). *Sharks and Rays of Australia.* CSIRO, Melbourne.

Neira FJ, Miskiewicz AG and Trinski T (1998). *Larvae of temperate Australian fishes, laboratory guide for larval fish identification.* University of Western Australia Press, Perth.

Shepherd SA and Thomas IM (eds) (1982). *Marine Invertebrates of Southern Australia. Part I.* Handbooks of the Flora and Fauna of South Australia. Government Printers, Adelaide. [environment, sessile animals, echinoderms]

Shepherd SA and Thomas IM (eds) (1989). *Marine Invertebrates of Southern Australia. Part II.* Handbooks of the Flora and Fauna of South Australia. Government Printers, Adelaide. [molluscs]

Shepherd SA and Davies M (eds) (1997). *Marine Invertebrates of Southern Australia. Part III.* Handbooks of the Flora and Fauna of South Australia. Government Printers, Adelaide. [worms, brachiopods, pycnogonids, ascidians]

Womersley HBS (1984). *The Marine Benthic Flora of Southern Australia. Part I.* Handbook of the Flora and Fauna of South Australia. Government Printer, Adelaide. [biogeography, ecology, seagrasses, green algae]

Womersley HBS (1987). *The Marine Benthic Flora of Southern Australia. Part II.* Handbook of the Flora and Fauna of South Australia. Government Printer, Adelaide. [red algae]

Womersley HBS (1988). *The Marine Benthic Flora of Southern Australia. Part IIIA.* Handbook of the Flora and Fauna of South Australia. Government Printer, Adelaide. [brown algae]

FISHERIES AND ECOLOGY

Clayton MN and King RJ (eds) (1981). *Marine Botany: An Australian Perspective.* Longman Cheshire, Melbourne.

DPIE (1995). *Marketing names for fish and seafood in Australia.* Dept of Primary Industries & Energy, Canberra.

Kailoa PJ, Williams MJ, Stewart PC, Reichelt RE, McNee A and Grieve C (1993). *Australian Fisheries Resources.* Bureau of Resource Sciences, Canberra.

Kingsford MJ and Battershill CN (eds) (1998). *Studying temperate marine environments: a handbook for ecologists.* Canterbury University Press, Christchurch.

Underwood AJ and Chapman MG (eds) (1995). *Coastal Marine Ecology in Temperate Australia.* New South Wales University Press, Sydney.

CHAPTER 1 OCEANOGRAPHY AND BIOGEOGRAPHY

Allan R, Lindesay J and Parker D (1996). *El Niño southern oscillation and climatic variability.* CSIRO Publishing, Melbourne.

Bouma WJ, Pearman GI, and Manning MR (1996). *Greenhouse: coping with climate change.* CSIRO Publishing, Melbourne.

Caputi N, Fletcher WJ, Pearce A, and Chubb CF (1996). Effect of the Leeuwin Current on the recruitment of fish and invertebrates along the Western Australian coast. *Marine and Freshwater Research* 47, 147–56.

Church JA and Craig PD (1998). Australia's shelf seas: diversity and complexity. In, *The Sea, Vol II.* AR Robinson and KH Brink (eds). John Wiley & Sons Inc, New York, pp 933–64.

Cresswell GR, and Legeckis R (1986). Eddies off south-eastern Australia. *Deep-Sea Research* 33, 1527–62.

Daly KL, and Smith OJ (1993). Physical-biological interactions influencing marine plankton production. *Annual Review of Ecology and Systematics* 24, 555–86.

Edyvane K (1996). The role of marine protected areas in temperate ecosystem management. In, *Developing Australia's representative system of marine protected areas: Criteria and guidelines for identification and selection.* Proceedings of a technical meeting held at South Australian Aquatic Sciences Centre, West Beach Adelaide, 22–23 April 1996. Thackway R (ed). Department of the Environment, Sport and Territories, Canberra, pp 53–67.

Griffin DA and Middleton JH (1991). Local and remote wind forcing of New South Wales inner shelf currents and sea level. *Journal of Physical Oceanography* 21, 304–22.

Grimes CB and Kingsford MJ (1996). How do estuarine and riverine plumes of different sizes influence fish larvae: do they enhance recruitment? *Marine and Freshwater Research* 47, 191–208.

Hutchins JB and Pearce AF (1994). Influence of the Leeuwin current on recruitment of tropical reef fishes at Rottnest Island, Western Australia. *Bulletin of Marine Science* 54, 245–55.

Kingsford MJ (1990). Linear oceanographic features: a focus for research on recruitment processes. *Australian Journal of Ecology* 15, 391–401.

Middleton JH, Cox D and Tate P (1996). The oceanography of the Sydney Region. *Marine Pollution Bulletin* 33, 124–31.

Pearce AF and Walker DI (1991). The Leeuwin Current: an influence on the coastal climate and marine life of Western Australia. *Journal of the Royal Society of Western Australia* 74, 1–140.

Poore GCB (1995). Biogeography and diversity of Australia's marine biota. In, *The State of the Marine Environment Report for Australia. Technical Annex: 1 The Marine Environment.* Zann LP and Kailola P (eds), Great Barrier Reef Marine Park Authority, Canberra, pp 75–84.

Randall JE (1981) Examples of antitropical and antiequatorial distribution of Indo-west Pacific fishes. *Pacific Science* 35, 197–209.

Short AD and Trenaman NL (1992). Wave climate of the Sydney region, an energetic and highly variable ocean regime. *Australian Journal of Marine and Freshwater Research* 43, 765–91.

CHAPTER 2 NEW SOUTH WALES

Andrew NL (1991). Changes in subtidal habitat following mass mortality of sea urchins in Botany Bay, New South Wales. *Australian Journal of Ecology* 16, 353–62.

Andrew NL and Underwood AJ (1989). Patterns of abundance in the sea urchin *Centrostephanus rodgersii* on the central coast of New South Wales, Australia. *Journal of Experimental Marine Biology and Ecology* 131, 61–80.

Andrew NL and Underwood AJ (1992). Association and abundance of sea urchins and abalone on shallow subtidal reefs in southern New South Wales. *Australian Journal of Marine and Freshwater Research* 43, 1547–59.

Davis AR, Roberts DE and Cummins SP (1997). Rapid invasion of sponge-dominated deep-reef by *Caulerpa scalpelliformis. Australian Journal of Ecology* 22, 146-150.

Harriott, VJ, Smith, SDA and Harrison, PL (1994). Patterns of coral community structure of subtropical reefs in the Solitary Islands Marine Reserve, Eastern

Australia. *Marine Ecology Progress Series* 109, 67–76.
Holbrook SJ, Kingsford MJ, Schmitt RJ and Stephens JS (1990). Spatial and temporal patterns in assemblages of temperate reef fish. *American Zoologist* 34, 463–75.
Jones GP and Andrew NL (1993). Temperate reefs and the scope of seascape ecology. In, *Proceedings of the Second International Temperate Reef Symposium, Auckland New Zealand.* Battershill CN, Schiel DR, Jones GP, Creese RG and MacDiarmid AB (eds), NIWA, Wellington, New Zealand, pp 63–77.
Roberts DE (1996). Patterns in subtidal marine assemblages associated with a deep-water sewage outfall. *Marine and Freshwater Research* 47, 1–9.
Roberts DE and Davis AR (1996). Patterns in sponge (Porifera) assemblages on temperate coastal reefs off Sydney, Australia. *Marine and Freshwater Research* 47, 897–906.
Underwood AJ, Kingsford MJ and Andrew NL (1991). Patterns in shallow subtidal marine assemblages along the coast of New South Wales. *Australian Journal of Ecology* 6, 231–49.

CHAPTER 3
VICTORIA

Anon (1993). *Marine and Coastal Special Investigation: Descriptive report.* Land Conservation Council, Melbourne.
O'Hara TD (1993). Echinoderms of Victoria. *Victorian Naturalist* 110, 149–53.
O'Toole M and Turner M (1990). *Down under at the Prom.* Field Naturalist Club of Victoria and Department of Conservation and Environment, Melbourne.
Phillips D, Handreck C, Bock P, Burn R, Smith B and Staples D (1984). *Coastal invertebrates of Victoria: An atlas of selected species.* Marine Research Group and Museum of Victoria, Melbourne.
Turner ML and Norman MD (1998). Fishes of Wilsons Promontory and Corner Inlet, Victoria: composition and biogeographic affinities. *Memoirs of the Museum of Victoria* 57, 143–65.
Wilson RS, Poore GCB and Gomon MF (1990). *Marine habitats at Wilsons Promontory and the Bunurong coast, Victoria: report on a survey,* 1982. Marine Science Laboratories Technical Report No 73.

CHAPTER 4
TASMANIA

Bruce BD, Green MA and Last PR (1998). Threatened fishes of the world: spotted handfish, *Brachionichthys hirsutus* (Lacepede*). Environmental Biology of Fishes* 52, 418.
Edgar GJ (1984) General features of the ecology and biogeography of Tasmanian rocky reef communities. *Papers and Proceedings of the Royal Society of Tasmania* 118, 173–86.
Edgar GJ and Barrett NS (1997). Short-term monitoring of biotic change in Tasmanian marine reserves. *Journal of Experimental Marine Biology and Ecology* 213, 261–79.
Edgar GJ, Moverley JS, Barrett NS, Peters D and Reed C (1997). The conservation-related benefits of a systematic marine biological sampling program: the Tasmanian bioregionalisation as a case study. *Biological Conservation* 79, 227–40.
Harris S, Brothers N, Coates F, Edgar GJ, Last PR, Richardson AMM, and Wells P (1993). The biological significance of a coastline in the roaring forties latitude. In, *Tasmanian Wilderness — World Heritage Values.* Smith SJ and Banks MR (eds). Royal Society of Tasmania, Hobart, pp 123–8.
Harris GP, Nilsson CS, Clementson LA and Thomas DP (1987). The water masses of the east coast of Tasmania: seasonal and interannual variability and the influence on phytoplankton biomass and productivity. *Australian Journal of Marine and Freshwater Research* 38, 569–90.
Last PR (1989). Nearshore habitats. In, *Is history enough? Past, present and future use of the resources of Tasman Peninsula.* Smith SJ (ed), Royal Society of Tasmania, Hobart, pp 71–80.

CHAPTER 5
SOUTH AUSTRALIA

Branden KL, Edgar GJ and Shepherd SA (1986). Reef fish populations of the Investigator Group, South Australia: a comparison of two census methods. *Transactions of the Royal Society of South Australia* 110, 60–76.
Edyvane K (1996). Marine environmental issues in South Australia. In, *State of the Marine Environment Report for Australia. Technical Annex III. State and Territory Issues.* Zann LP and Sutton D (eds), Great Barrier Reef Marine Park Authority, Canberra, pp 61–88.
Lewis RK, Edyvane K and Newland N (eds) (1998). *Description, Use and Management of South Australia's Marine and Estuarine Environment.* Government of South Australia.
Shepherd SA (1981). Ecological strategies of a deep water red algal community. *Botanica Marina* 24, 457–63.
Shepherd SA and Sprigg RC (1976). Substrate, sediments and subtidal ecology of Gulf San Vincent and Investigator Strait. In, *Natural History of the Adelaide Region.* Twidale CR, Tyler MJ and Webb BP (eds). Royal Society of South Australia, pp 161–74.
Shepherd SA and Womersley HBS (1970). The sublittoral ecology of West Island, South Australia. I. Environmental factors and algal ecology. *Transactions of the Royal Society of South Australia* 94, 105–38.
Shepherd SA and Womersley HBS 1971. Pearson Island Expedition 1969. I. The subtidal ecology of benthic algae. *Transactions of the Royal Society of South Australia* 95, 155–67.
Shepherd SA and Womersley HBS (1976). The subtidal algal and seagrass ecology of St. Francis Island, South Australia. *Transactions of the Royal Society of South Australia* 100, 177–91.
Shepherd SA and Womersley HBS (1981). The algal and seagrass ecology of Waterloo Bay, South Australia. *Aquatic Botany* 11, 305–71.
Whitley G and Allan J (1958). *The Sea-horse and its Relatives.* Georgian House, Melbourne.
Womersley HBS and King RJ (1990). The ecology of temperate rocky shores. In, *Biology of Marine Plants,* MN Clayton and RJ King (eds). Longman Cheshire, Melbourne, pp 266–95.

CHAPTER 6
WESTERN AUSTRALIA

Berry PF and Playford PE (1992). Territoriality in a subtropical kyphosid fish associated with macroalgal polygons on reef platforms at Rottnest Island, Western Australia. *Journal of the Royal Society of Western Australia* 75, 67–73.
Hatcher A (1989). Variation in the components of benthic community structure in a coastal lagoon as a function of spatial scale. *Australian Journal of Marine and Freshwater Research* 40, 79–96.
Hatcher BG, Kirkman H and Wood WF (1987). Growth of the kelp *Ecklonia radiata* near the northern limit of its range in Western Australia. *Marine Biology* 95, 63–73.
Kendrick GA (1994). Effects of settlement density and adult canopy on survival of recruits of *Sargassum* spp. (Sargassaceae, Phaeophyta) *Marine Ecology Progress Series* 103, 129–40.
Kendrick GA and Walker DI (1994). Role of recruitment in structuring beds of *Sargassum* spp. (Phaeophyta) at Rottnest Island, Western Australia. *Journal of Phycology* 30, 200–8.
Kendrick GA and Walker DI (1995). Dispersal of propagules of *Sargassum* spp. (Sargassaceae, Phaeophyta): observations of local patterns of dispersal and possible consequences for recruitment and population structure. *Journal of Experimental Marine Biology and Ecology* 192, 273–88.
Kirkman H (1989). Growth, density and biomass of *Ecklonia radiata* at different depths and growth under artificial shading off Perth, Western Australia. *Australian Journal of Marine and Freshwater Research* 40, 169–97.
Lenanton RCJ, Robertson AI and Hansen JA (1982). Nearshore accumulations of detached macrophytes as nursery areas for fish. *Marine Ecology Progress Series* 9, 51–7.
Phillips JC, Kendrick GA and Lavery PJ (1997). A test of a functional group approach to detecting shifts in macroalgal communities along a disturbance gradient. *Marine Ecology Progress Series* 153, 125–38.
Wells FE (ed) (1997). *Proceedings of the 7th International Marine Biological Workshop: The marine flora and fauna of the Houtman Abrolhos Islands, Western Australia.* Western Australian Museum, Perth.
Wells FE, Walker DI, Kirkman H and Lethbridge R (eds) (1991). *Proceedings of the 3rd International Marine Biological Workshop: The flora and fauna of Albany, Western Australia.* Western Australian Museum, Perth.
Wells FE, Walker DI, Kirkman H and Lethbridge R (eds) (1993). *Proceedings of the 5th International Marine Biological Workshop: the flora and fauna of Rottnest Island, Western Australia.* Western Australian Museum, Perth.

CHAPTER 7
KELP FORESTS

Andrew NL (1993). Spatial heterogeneity, sea urchin grazing and habitat structure on reefs in temperate Australia. *Ecology* 74, 292–302.
Clayton MN (1990). The adaptive significance of life history characters in selected orders of marine brown macroalgae. *Australian Journal of Ecology* 15, 439–52.
Jennings J and Steinberg PD (1997). Phlorotannins versus other factors affecting epiphyte abundance on the kelp *Ecklonia radiata. Oecologia* 109, 461–73.

Kendrick GA (1994). Effects of settlement density and adult canopy on survival of recruits of *Sargassum* spp. (Sargassaceae, Phaeophyta). *Marine Ecology Progress Series* 103, 129–40.

Kendrick GA and Walker DI (1995). Dispersal of propagules of *Sargassum* spp. (Sargassaceae, Phaeophyta): observations of local patterns of dispersal and possible consequences for recruitment and population structure. *Journal of Experimental Marine Biology and Ecology* 192, 273–88.

Kennelly SJ (1987). Physical disturbances in an Australian kelp community. I. Temporal effects. *Marine Ecology Progress Series* 40, 145–53.

Kennelly SJ and Underwood AJ (1992). Fluctuations in the distributions and abundances of species in sublittoral kelp forests in New South Wales. *Australian Journal of Ecology* 17, 367–82.

Kirkman H (1981). The first year in the life history and the survival of the juvenile marine macrophyte, *Ecklonia radiata* (Turn.) J. Agardh. *Journal of Experimental Marine Biology and Ecology* 55, 243–54.

Kirkman H. (1984). Standing stock and production of *Ecklonia radiata* (C.Ag.) J. Agardh. *Journal of Experimental Marine Biology and Ecology* 76, 119–30.

CHAPTERS 8 AND 9
BLACKLIP AND GREENLIP ABALONE

Much of the scientific information on abalone has been gathered in the volumes edited by Shepherd et al. cited below. More recent references are also provided.

Andrew NL, Worthington DG and Brett PA (1997). Size structure and growth of individuals suggest high exploitation rates within the fishery for blacklip abalone *Haliotis rubra*, in NSW, Australia. *Molluscan Research* 18, 275–87.

Andrew NL, Worthington DG, Brett PA, Bentley N, Chick R and Blount C (1998). *Interactions between the abalone fishery and sea urchins in New South Wales.* Final report to the Fisheries Research and Development Corporation.

Shepherd SA, Day RW and Butler AJ (eds) (1995). Progress in abalone fisheries research. *Marine and Freshwater Research* Vol. 46.

Shepherd SA, McShane PE and Wells FE (eds) (1997). Australasian abalone. *Molluscan Research* Vol. 18.

Shepherd SA, Tegner, MJ and Guzmán del Próo SA (eds) (1992). *Abalone of the world: biology, fisheries and culture.* Blackwells, Oxford.

Worthington DG and Andrew NL (1998). Small scale variation in demography and its implications for alternative size limits in the fishery for abalone in NSW, Australia. *Special Publication of the Canadian Journal of Fisheries and Aquatic Sciences* 125, 341–8.

CHAPTER 10
OCTOPUSES

Boyle PR (ed) (1983). *Cephalopod Life Cycles. Volume 1. Species Accounts.* Academic Press, London.

Boyle PR (ed) (1987). *Cephalopod Life Cycles. Volume 2. Comparative Reviews.* Academic Press, London.

Hanlon RT and Messenger JB (1996). *Cephalopod behaviour.* Cambridge University Press, Cambridge.

Lu CC and Dunning MC (1998). Subclass Coleoidea. In, *Mollusca: The Southern Synthesis. Fauna of Australia.* Vol. 5. Beesley PL, Ross GJB and Wells A (eds). CSIRO Publishing, Melbourne, pp 499–563.

Norman MD and Reid A (1998). *A guide to cephalopods (octopuses, cuttlefishes and squids) of Australia and South Asia.* Gould League of Victoria and CSIRO, Melbourne.

CHAPTER 11
JELLYFISH

Arai MN (1997). *A Functional Biology of Scyphozoa.* Chapman and Hall, London.

Calder DR (1982). Life history of the cannonball jellyfish, *Stomolophus meleagris* L. Agassiz, 1860 (Scyphozoa, Rhizostomida). *Biological Bulletin* 162, 149–62.

Fancett MS (1986). Species Composition and Abundance of Scyphomedusae in Port Phillip Bay, Victoria. *Australian Journal of Marine and Freshwater Research* 37, 379–84.

Fancett MS (1988). Diet and prey selectivity of scyphomedusae from Port Phillip Bay, Australia. *Marine Biology* 98, 503–9.

Rippingale RJ and Kelly SJ (1995). Reproduction and survival *of Phyllorhiza punctata* (Cnidaria: Rhizostomeae) in a seasonally fluctuating salinity regime in Western Australia. *Marine and Freshwater Research* 46, 1145–51.

Williamson JA, Fenner PJ, Burnett JW and Rifkin JF (1996). *Venomous and Poisonous Marine Animals: A Medical and Biological Handbook.* University of New South Wales Press, Australia.

CHAPTER 12
SOUTHERN ROCK LOBSTERS

Booth JD (1994). *Jasus edwardsii* larval recruitment off the east coast of New Zealand. *Crustaceana* 66, 295–317.

Booth JD, Carruthers AD, Bolt CD and Stewart RA (1991). Measuring depth of settlement in the red rock lobster, *Jasus edwardsii. New Zealand Journal of Marine and Freshwater Research* 25, 123–32.

Frusher SD, Kennedy RB and Gibson ID (1998). Preliminary estimates of exploitation rates in the Tasmanian rock lobster (*Jasus edwardsii*) fishery using change-in-ratio and index-removal techniques with tag-recapture data. *Special Publication of the Canadian Journal of Fisheries and Aquatic Sciences* 125, 63–71.

MacDiarmid AB (1989). Size at onset of maturity and size-dependent reproductive output of female and male spiny lobsters *Jasus edwardsii* (Hutton Decapoda, Palinuridae) in northern New Zealand. *Journal of Experimental Marine Biology and Ecology* 127, 229–43.

MacDiarmid AB (1989). Moulting and reproduction of the spiny lobsters *Jasus edwardsii* (Decapoda, Palinuridae) in northern New Zealand. *Marine Biology* 103, 303–10.

CHAPTER 13
EASTERN ROCK LOBSTERS

Booth JD (1986). Recruitment of packhorse lobster *Jasus verreauxi* in New Zealand. *Canadian Journal of Fisheries and Aquatic Sciences* 43, 2212–20.

Booth JD and Phillips BF (1994). Early life history of spiny lobsters. *Crustaceana* 66, 271–94.

Brasher DJ, Ovenden JD, Booth JD and White RWG (1992). Genetic subdivision of Australian and New Zealand populations of *Jasus verreauxi* (Decapoda: Palinuridae) — preliminary evidence from mitochondrial genome. *New Zealand Journal of Marine and Freshwater Research* 26, 53–8.

Montgomery SS (1992). Sizes at first maturity and at onset of breeding in female *Jasus verreauxi* (Decapoda: Palinuridae) from New South Wales, Australia. *Australian Journal of Marine and Freshwater Research* 43, 1373–9.

Montgomery SS (1995). Patterns in landings and size composition of *Jasus verreauxi* (H. Milne Edwards, 1851) (Decapoda, Palinuridae), in waters off New South Wales, Australia. *Crustaceana* 68, 257–66.

CHAPTER 14
WESTERN ROCK LOBSTERS

Anon (1997). Fishing for rock lobsters: 1997/98 season. *Fish for the future brochure, September 1997,* Fisheries Department of Western Australia, Perth.

Caputi N and Brown RS (1993). The effect of environment on puerulus settlement of the Western Rock Lobster (*Panulirus cygnus*) in Western Australia. *Fisheries and Oceanography* 2, 1–10.

Caputi N, Brown RS and Chubb CF (1995). Regional prediction of the Western Rock Lobster, *Panulirus cygnus,* commercial catch in Western Australia. *Crustaceana* 68, 245–56.

Chubb CF, Barker EH and Dibden CJ (1994). The Big Bank region of the limited entry fishery for the Western Rock Lobster, *Panulirus cygnus. Western Australian Marine Research Laboratories Fisheries Research Report 101,* Fisheries Department of Western Australia, Perth.

Gray H (1992). The Western Rock Lobster *Panulirus cygnus. Book 1: A Natural History,* Westralian Books, Geraldton.

Melville-Smith R, Chubb CF, Caputi N, Cheng YW, Christianopoulos D and Rossbach M (1998). *Fishery independent survey of the breeding stock and migration of the Western Rock Lobster* (*Panulirus cygnus*). Final report to The Fisheries Research and Development Corporation, Canberra.

Phillips BF, Cobb JS and Kittaka J (eds) (1994). *Spiny Lobster Management.* Fishing News Books, London.

Phillip BF and Pearce AF (1997). Spiny lobster recruitment off Western Australia. *Bulletin of Marine Science* 61, 21–41.

Rossbach M, Chubb CF, Melville-Smith R and Cheng YW (1997). *Mortality, growth and movement of the Western Rock Lobster.* Final report to the Fisheries Research and Development Corporation, Canberra.

CHAPTER 15
SEA URCHINS

Andrew NL (1989). Contrasting ecological implications of food limitation in sea urchins and herbivorous gastropods. *Marine Ecology Progress Series* 51, 189–93.

Andrew NL (l994). Survival of kelp adjacent to areas grazed by sea urchins in New South Wales,

Australia. *Australian Journal of Ecology* 16, 353–62.
Byrne M, Andrew NL, Worthington DG and Brett PA (1998). The influence of latitude and habitat on reproduction in the sea urchin *Centrostephanus rodgersii* in New South Wales, Australia. *Marine Biology* 132, 305–18.
Constable AJ (1993). The role of sutures in shrinking of the test in *Heliocidaris erythrogramma* (Echinoidea: Echinometridae). *Marine Biology* 117, 423–30.
Dix TG (1977). Reproduction in Tasmanian populations of *Heliocidaris erythrogramma* (Echinodermata: Echinometridae). *Australian Journal of Marine and Freshwater Research* 28, 509–20.
Fletcher WJ (1987). Interactions among subtidal Australian sea urchins, gastropods and algae: effects of experimental removals. *Ecological Monographs* 57, 89–109.
Growns JE and Ritz DA (1994). Colour variation in southern Tasmanian populations of *Heliocidaris erythrogramma* (Echinometridae: Echinoidea). *Australian Journal of Marine and Freshwater Research* 45, 233–42.
Laegdsgaard P, Byrne M and Anderson DT (1991). Reproduction of sympatric populations of *Heliocidaris erythrogramma* and *H. tuberculata* (Echinoidea) in New South Wales. *Marine Biology* 110, 359–74.
Williams DHC and DT Anderson (1975). The reproductive system, embryonic development, larval development and metamorphosis of the sea urchin *Heliocidaris erythrogramma* (Val.) (Echinoidea, Echinometridae). *Australian Journal of Zoology* 23, 371–403.

CHAPTER 16
SESSILE ANIMALS

Ayre DJ (1990). Population subdivision in Australian temperate marine invertebrates: larval connections versus historical factors. *Australian Journal of Ecology* 15, 403–12.
Buss LW (1990). Competition within and between encrusting clonal invertebrates. *Trends in Ecology and Evolution* 5, 352–6.
Connell JH and Keough MJ (1985). Disturbance and patch dynamics of subtidal marine animals on hard substrata. In, *The ecology of natural disturbance and patch dynamics*. STA Pickett and PS White (eds), Academic Press, New York, pp 125–52.
Davis AR (1996). Association among ascidians — facilitation of recruitment in *Pyura spinifera*. *Marine Biology* 126, 35–41.
Hughes TP, Ayre D and Connell JH (1992). The evolutionary ecology of corals. *Trends in Ecology and Evolution* 7, 292–5.
Hughes TP and Jackson JBC (1980). Do corals lie about their age? Some demographic consequences of partial mortality, fission and fusion. *Science* 209, 713–5.
Jackson JBC Buss LW and Cook RE (1985). *Population Biology and Evolution of Clonal Organisms*. Yale University Press, New Haven.
Keough MJ (1984). Dynamics of the epifauna of the bivalve *Pinna bicolor*: interactions among recruitment, predation, and competition. *Ecology* 65, 677–88.
Keough MJ and Butler AJ (1995). Temperate subtidal hard substrata. In, *State of the Marine Environment Report for Australia. Technical Annex I. The Marine Environment.* Zann L and Kailola P (eds), Great Barrier Reef Marine Park Authority, Canberra, pp 37–52.
Klemke JE and Keough MJ (1991). Relationships between the nudibranch *Madrella sanguinea* and its broadly-distributed prey, the bryozoan *Mucropetraliella ellerii*. *Journal of Molluscan Studies* 57, 23–33.

CHAPTER 17
REEF SHARKS AND RAYS

Bass AJ, D'Aubrey JD and Kistnasamy N (1975). *Sharks of the East Coast of Southern Africa. IV. The Families Odontaspididae, Scapanorhynchidae, Isuridae, Cetorhinidae, Alopiidae, Orectolobidae and Rhiniodontidae. In-vestigational Report No. 39*, The Oceanographic Research Institute, Durban, South Africa.
Campagno LJV (1984). *Sharks of the World. FAO Fisheries Synposis, Volume 4.* United Nations, Food and Agricultural Organisation, Rome.
Dayton PK, Tegner MJ, Edwards PB and Riser KL (1998). Sliding baselines, ghosts and reduced expectations in kelp forest communities. *Ecological Applications* 8, 309–22.
Lincoln Smith MP, Hair CA and Bell JD (1992). *Jervis Bay Marine Ecological Study, Project 4: Fish Associated with Natural Rocky Reefs and Artificial Breakwaters.* New South Wales Fisheries Research Institute, Cronulla, New South Wales.
McLaughlin RH and O'Gower AK (1971). Life history and underwater studies of a heterodontid shark. *Ecological Monographs* 41, 271–89.
O'Gower AK and Nash AR (1978). Dispersion of Port Jackson shark in Australian waters. In, *Sensory Biology of Sharks, Skates and Rays.* Hodgson ES and Mathewson RF (eds), US Department of Navy, Office of Naval Research, Arlington, pp 529–44.
Pollard DA, Lincoln Smith MP and Smith AK (1996). The biology and conservation status of the grey nurse shark (*Carcharias taurus* Rafinesque 1810) in New South Wales, Australia. *Aquatic Conservation: Marine and Freshwater Ecosystems* 6, 1–20.
Roughley TC (1955). *Fish and Fisheries of Australia.* 3rd ed. Angus and Robertson, Sydney.
Whitley GP (1983). *Sharks of Australia.* Revision edited by MP Lincoln Smith. Jack Pollard Publications, Sydney.

CHAPTER 18
SNAPPER AND KINGFISH

Ferrell DJ and Sumpton W (1997). *Assessment of the fishery for snapper (*Pagrus auratus*) in Queensland and New South Wales.* Final Report to Fisheries Research and Development Corporation.
Francis MP and Pankhurst NW (1988). Juvenile sex inversion in the New Zealand Snapper *Chrysophrys auratus* (Bloch and Schneider, 1801) (Sparidae). *Australian Journal of Marine and Freshwater Research* 39, 625–31.
Gillanders BM, Ferrell DJ and Andrew NL (1997). *Determination of ageing in Kingfish (*Seriola lalandi*) in New South Wales.* Final Report to Fisheries Research and Development Corporation.
Holdsworth J (1995). Drifting buoy offers clues on juvenile kingfish. *Seafood New Zealand 3*, page19.
Schmitt RJ and Strand SW (1982). Cooperative foraging by yellowtail, *Seriola lalandi* (Carangidae), on two species of fish prey. *Copeia* 82, 714–17.
Smale MJ (1986). The feeding habits of six pelagic and predatory teleosts in eastern Cape coastal waters (South Africa). *Journal of Zoology, London* 1, 357–409.
Smith A, Pepperell J, Diplock J and Dixon P (1991). Study suggests NSW kingfish one stock. *Australian Fisheries* March, 34–6.

CHAPTER 19
TERRITORIAL DAMSELFISHES

Dove SG and Kingsford MJ (1998). The use of otoliths and eye lenses for measuring trace metal incorporation in fish; a biogeographic study. *Marine Biology* 130, 377–88.
Holbrook SJ, Kingsford MJ, Schmitt RJ and Stephens JS (1994). Spatial patterns of marine reef fish assemblages. *American Zoologist* 34, 463–75.
Jones GP and Norman MD (1986). Feeding selectivity in relation to territory size in a herbivorous reef fish. *Oecologia* 68, 549–56.
Moran MJ and Sale PF (1977). Seasonal variation in territorial response, and other aspects of the ecology of the Australian temperate pomacentrid fish *Parma microlepis*. *Marine Biology* 39, 121–8.
Norman MD and Jones GP (1984). Determinants of territory size in the pomacentrid reef fish, *Parma victoriae*. *Oecologia* 61, 60–9.
Tzioumis V and Kingsford MJ (1995). Periodicity of spawning of two temperate damselfishes: *Parma microlepis* and *Chromis dispilus*. *Bulletin of Marine Science* 57, 596–609.
Tzioumis V and Kingsford MJ (1999). Reproductive biology and growth of the temperate damselfish *Parma microlepis*. *Copeia* 2, 348–61.

CHAPTER 20
MORWONG

Bell JD (1979). Observations on the diet of Red Morwong, *Cheilodactylus fuscus* Castelnau (Pisces: Cheilodactylidae). *Australian Journal of Marine and Freshwater Research* 30, 129–33.
Cappo M (1995). Population biology of the temperate reef fish *Cheilodactylus nigripes* in an artificial reef environment. *Transactions of the Royal Society of South Australia* 119, 112–22.
Leum L and Choat J (1980). Density and distribution patterns of the temperate marine fish *Cheilodactylus spectabilis* (Cheilodactylidae) in a reef environment. *Marine Biology* 57, 327–37.
Lincoln Smith MP, Bell JD, Pollard DA and Russell BC (1989). Catch and effort of competition spearfishermen in south eastern Australia. *Fisheries Research* 8, 45–61.
McCormick MI (1989a). Spatio-temporal patterns in the abundance and population structure of a large temperate reef fish. *Marine Ecology Progress Series* 53, 215–25.
McCormick MI (1989b). Reproductive ecology of the temperate reef fish *Cheilodactylus spectabilis* (Pisces: Cheilodactylidae). *Marine Ecology Progress Series* 55, 113–20.
McCormick MI (1998). Ontogeny of diet shifts by a micro-carnivorous fish, *Cheilodactylus spectabilis*:

relationship between feeding mechanics, microhabitat selection and growth. *Marine Biology* 132, 9–20.

Schroeder A, Lowry M and Suthers I (1994). Sexual dimorphism in the Red Morwong (*Cheilodactylus fuscus*). *Australian Journal of Marine and Freshwater Research* 45, 1173–80.

Vooren CM (1978). Post-larvae and juveniles of the tarakihi (Pisces: Cheilodactylidae) in New Zealand. *New Zealand Journal of Marine and Freshwater Research* 6, 601–18.

CHAPTER 21
THE WRASSES

Edgar GJ and Barrett NS (1997). Short-term monitoring of biotic change in Tasmanian marine reserves. *Journal of Experimental and Marine Biology and Ecology* 213, 261–79.

Jones GP (1988). Ecology of rocky reef fish of north-eastern New Zealand. *New Zealand Journal of Marine and Freshwater Research* 22, 445–62.

McPherson GR (1977). Sex change in the wrasse *Pseudolabrus gymnogenis* (Labridae). *Australian Zoologist* 19, 185–200.

Robertson DR (1972). Social control of sex reversal in a coral reef fish. *Science* 177, 1007–9.

Russell BC (1988). Revision of the labrid fish genus *Pseudolabrus* and allied genera. *Records of the Australian Museum*, Supplement 9, page 72.

Shepherd SA and Hobbs LJ (1985). Age and growth of the blue-throated wrasse *Pseudolabrus tetricus. Transactions of the Royal Society of Australia* 109, 177–8.

CHAPTER 22
BLUE GROPER

Gillanders BM (1995a). Feeding ecology of the temperate marine fish, *Achoerodus viridis*, (Labridae): Size, seasonal and site-specific differences. *Marine and Freshwater Research* 46, 1009–20.

Gillanders BM (1995b). Reproductive biology of the protogynous hermaphrodite *Achoerodus viridis* (Labridae) from south-eastern Australia. *Marine and Freshwater Research* 46, 999–1008.

Gillanders BM (1997a). Comparison of growth rates between estuarine and coastal reef populations of *Achoerodus viridis* (Pisces: Labridae). *Marine Ecology Progress Series* 146, 283–7.

Gillanders BM (1997b). Patterns of abundance and size structure in the blue groper *Achoerodus viridis* (Pisces: Labridae): evidence of links between estuaries and coastal reefs. *Environmental Biology of Fishes* 49, 153–73.

Gillanders BM and Kingsford MJ (1996). Elements in otoliths may elucidate the contribution of estuarine recruitment to sustaining coastal reef populations of a temperate reef fish. *Marine Ecology Progress Series* 141, 13–20.

Gomon MF (1997). Relationships of fishes of the labrid tribe hypsigenyini. *Bulletin of Marine Science* 60, 789–871.

McNeill SE, Worthington DG, Ferrell DJ and Bell JD (1992). Consistently outstanding recruitment of five species of fish to a seagrass bed in Botany Bay, N.S.W. *Australian Journal of Ecology* 15, 360–7.

Worthington DG, Ferrell DJ, McNeill SE and Bell JD (1992). Growth of four species of juvenile fish associated with the seagrass, *Zostera capricorni*, in Botany Bay, New South Wales. *Australian Journal of Marine and Freshwater Research* 43, 1189–98.

CHAPTER 23
LEATHERJACKETS

Bell JD, Burchmore JJ and Pollard DA (1978). Feeding ecology of three sympatric species of leatherjackets (Pisces: Monacanthidae) from a *Posidonia* seagrass habitat in New South Wales. *Australian Journal of Marine and Freshwater Research* 29, 631–43.

Conacher MJ, Lanzing WJR and Larkum AWD (1979). Ecology of Botany Bay. II. Aspects of the feeding ecology of the Fanbellied Leatherjacket, *Monacanthus chinensis* (Pisces: Monacanthidae), in *Posidonia australis* seagrass beds in Quibray Bay, Botany Bay, New South Wales. *Australian Journal of Marine and Freshwater Research* 30, 387–400.

Grove-Jones R and Burnell AF (1991). *Fisheries biology of the Ocean Jacket (Monacanthidae: Nelusetta ayraudi) in the eastern waters of the Great Australian Bight, South Australia.* South Australian Department of Fisheries, Adelaide.

Kingsford MJ and Milicich M (1987). Presettlement phase of *Parika scaber* (Pisces: Monacanthidae), a temperate reef fish. *Marine Ecology Progress Series* 36, 65–79.

Russ G (1983). Temporal abundance of juveniles of a monacanthid fish (*Penicipelta vittiger* (Castelnau)) in southern Port Phillip Bay. *Victorian Naturalist* 100, 4–6.

CHAPTER 24
HERBIVOROUS FISHES

Anderson TA (1991). Mechanisms of digestion in the marine herbivore, the luderick, *Girella* (Quoy and Gaimard). *Journal of Fish Biology* 39, 535–47.

Andrew NL and Jones GP (1990). Patch formation by herbivorous fish in a temperate Australian kelp forest. *Oecologia* 85, 57–68.

Bell JD, Burchmore JJ and Pollard DA (1980). The food and feeding habits of the rock blackfish, *Girella elevata* Macleay (Pisces: Girellidae), from the Sydney region, New South Wales. *Australian Zoologist* 20, 391–407.

Berry PF and Playford PE (1992). Territoriality in a subtropical kyphosid fish associated with macroalgal polygons on reef platforms at Rottnest Island, Western Australia. *Journal of the Royal Society of Western Australia* 75, 67–73.

Choat JH and Clements KD (1992). Diet in odacid and aplodactylid fishes from Australia and New Zealand. *Australian Journal of Marine and Freshwater Research* 43, 1451–9.

Clements KD (1991). Endosymbiotic communities of two herbivorous labroid fishes, *Odax cyanomelas* and *O. pullus. Marine Biology* 109, 223–9.

Clements KD and Choat JH (1997). Comparison of herbivory in the closely-related marine fish genera *Girella* and *Kyphosus. Marine Biology* 127, 579–86.

Jones GP (1992). Interactions between herbivorous fishes and macro-algae on a temperate rocky reef. *Journal of Experimental Marine Biology and Ecology* 159, 217–35.

Rimmer DW (1986). Changes in the diet and the development of microbial digestion in juvenile buffalo bream, *Kyphosus cornelii. Marine Biology* 92, 443–8.

CHAPTER 25
PLANKTIVOROUS FISHES

Bray RN, Miller AC and Geesey GG (1981). The fish connection: a trophic link between planktonic and rocky reef communities? *Science* 214, 204–5.

Glasby TM and Kingsford MJ (1994). *Atypichthys strigatus* (Pisces: Scorpididae): an opportunistic planktivore that responds to benthic disturbances and cleans other fishes. *Australian Journal of Ecology* 19, 385–94.

Hobsen ES and Chess JR (1976). Trophic interactions among fishes and zooplankters near shore at Santa Catalina island, California. *Fishery Bulletin* 74, 567–99.

Kingsford MJ (1989). Distribution patterns of planktivorous reef fish along the coast of northeastern New Zealand. *Marine Ecology Progress Series* 54, 13–24.

Kingsford MJ and MacDiarmid AB (1988). Interrelations between planktivorous reef fish and zooplankton in temperate waters. *Marine Ecology Progress Series* 48, 103–17.

Sale PF (1991). *The ecology of fishes on coral reefs.* Academic Press, San Diego.

CHAPTER 26
SEALS AND SEA LIONS

Bryden M, Marsh H and Shaughnessy P (1998). *Dugongs, Whales, Dolphins and Seals; a Guide to the Sea Mammals of Australasia*, Allen and Unwin, Sydney.

Gales NJ, Coughran DK, and Queale LF (1992). Records of Subantarctic fur seals *Arctocephalus tropicalis* in Australia. *Australian Mammalogy* 15, 135–8.

Gales NJ, Shaughnessy PD and Dennis TE (1994). Distribution, abundance and breeding cycle of the Australian sea lion *Neophoca cinerea* (Mammalia: Pinnipedia). *Journal of Zoology, London* 23, 353–70.

Gales R and Pemberton D (1994). Diet of the Australian fur seal in Tasmania. *Australian Journal of Marine and Freshwater Research* 45, 653–64, [corrigendum] 1367.

Riedman M (1990). *The Pinnipeds: Seals, Sea Lions and Walrus.* University of California Press, Berkeley, California.

Shaughnessy PD and Davenport SR (1996). Underwater videographic observations and incidental mortality of fur seals around fishing equipment in south-eastern Australia. *Marine and Freshwater Research* 47, 553–6.

Shaughnessy PD, Gales NJ, Dennis TE and Goldsworthy SD (1994). Distribution and abundance of New Zealand fur seals *Arctocephalus forsteri* in South Australia and Western Australia. *Wildlife Research* 21, 667–95.

Slip DJ, Hindell MA and Burton HR (1994). Diving behavior of southern elephant seals from Macquarie Island: an overview. In, *Elephant seals: population ecology, behavior, and physiology.* Le Boeuf BJ and Laws RM (eds), University of California Press, Berkeley, California, pp 253–70.

Warneke RM (1995). Family Otariidae and Family Phocidae. In, *Mammals of Victoria; Distribution, Ecology and Conservation.* Menkhorst PW (ed), Melbourne, pp 244–56.

Index

Scientific names for species, genera, families and phyla are given in the index on page 236. They may also be found in the text with the first use of the common name. Page numbers in **bold** indicate photographs.

GENERAL INDEX

GENERAL INDEX OF SCIENTIFIC NAMES FOR SPECIES, GENERA AND FAMILIES

Common names are given in parentheses where appropriate. This index identifies only the pages where the scientific name is used; a complete list of references to species with common names is given in the General Index under their common name. Page numbers in **bold** indicate photographs.

gibbosus (Crested Morwong) 175
nigripes (Magpie Perch) 23, 34, 43, 175
rubrolabiatus (Red-lipped Morwong) 175
spectabilis (Banded Morwong) 9, 36, 175
vestitus (Magpie Morwong) 175
Chironex fleckeri (Tropical Box Jellyfish) 101
Chlorophyta 62
Choerodon rubescens (Baldchin Groper) 124, 187
Chondrilla 66
Chordata 140
Chromis 212
hypsilepis (One-spot Puller) 216
Cirrhitidae 173
Claudea 42
Clinidae 25
Cnemidocarpa pedata 19, 143, 144-45, **145**
Cnidaria 140
Cnidoglanus macrocephalus 56
Codium 42
dimorphum 36
Colurdontis paxmani (Paxman's Leatherjacket) 199
Comanthus 43
Conocladus australis **33**
Corallina 19, 44, 62, 66, 70
Coris 181
auricularis (Western King Wrasse) 186
picta (Combfish) 7, 9, 181
sandageri (Eastern King Wrasse) 185
Coscinaraea mcneilli 13
Coscinasterias muricata (Eleven-armed Seastar) 26, 82
Crassostrea gigas (Pacific Oyster) 37
Crinodus lophodon (Rock Cale) 16, 171, 205
Crustacea 140
Cyanea capillata (Lion's Mane Jellyfish) 99
Cymbastela concentrica 18
Cystophora 13, 23, 27, 36, 42, 48, 49, 53, 57, 61, 66, 71, 74
brownii 42, 48
congesta 42
expansa 48
gracilis 42
grevillei 42, 45
intermedia 44, 45
monilifera **63**
moniliformis 48
platylobium 42, 45
racemosa 45
retorta 45
retroflexa 42
sonderi 45
subfarcinata 45, 48
torulosa 37
Dactylophora nigricans (Dusky Morwong) 43, 174
Darwinella 66
Dascyllus aruanus (Humbug Damselfish) 170, 216
Dasyatidae 151, 157
Dasyatis
brevicaudata (Smooth Stingray) 157
thetidis (Black Stingray) 157
Dasycladus densus 42
Delisea 23, 67
pulchra 19, **65**, 70
Diadema palmeri **129**
Dicathais orbita (Cartrut Shell) 36
Dictymenia sonderi 66
Dictyota 62
dichotoma 19
Didemnum 66
Dinolestes lewini (Longfin Pike) 212
Diodon nicthemerus (Globe Fish) 25
Dissostichus eleginoides (Patagonian Toothfish) 225
Durvillaea potatorum (Bull Kelp) 9, 27, 31, 42, 128
Durvillaeales 61
Echinaster arcystatus 26
Echinodermata 127–35
Echinogorgia 48
Echinometra matheii (Tropical Sea Urchin) 56
Ecklonia radiata (Common Kelp) 13, 23, 36, 42, 49, 52, 61, 74, 116, 206
Electrona 223
Emmelichthys nitidus (Redbait) 224
Engraulis australis (Anchovy) 212
Enoplosus armatus (Old Wife) 23
Entacmaea quadricolor 168
Epinephelides armatus (Breaksea Cod) 124
Epinephelus rivulatus (Chinaman Cod) 124
Eubalaena australis (Southern Right Whale) 44
Eubalichthys 197
bucephalus (Black Reef Leatherjacket) 197
caeruleoguttatus (Blue-spotted Leatherjacket) 199
cyanoura (Blue-tailed Leatherjacket) 196
gunnii (Gunn's Leatherjacket) 196
mosaicus (Mosaic Leatherjacket) 101, 197
quadrispinis (Fourspine Leatherjacket) 197
Eudyptula minor (Little Penguin) 222
Euplexaura **48**
Euprymna tasmanica (Southern Dumpling Squid) 95
Filicampus tigris (Tiger Pipefish) 48
Forsterygion multiradiatum (Many-rayed Threefin) 34
Fromia polypora 26
Fucales 36, 61
Furgaleus macki (Whiskery Shark) 124
Gelidium 44
Genypterus tigerinus (Rock Ling) 25
Gigartina 23
Ginglymostomatidae 154
Girella 205
elevata (Eastern Rock Blackfish) 205
tephraeops (Western Rock Blackfish) 205
tricuspidata (Luderick) 205
zebra (Zebra Fish) 205
Girellidae 203
Glaphrymenia 42
Glaucosoma hebraicum (West Australian Dhufish) 124
Gnathanacanthus goetzeei (Red Velvetfish) 36
Gobiesocidae 82
Goniocidaris tubaria 44
Gracilaria 62
Grimpella thaumastocheir (Velvet Octopus) 92
Gymnoscopelus 224
Gymnothorax 171, 216
Halichoeres 181
brownfieldi (Brownfield's Wrasse) 184
Haliotis 25, 43
laevigata (Greenlip Abalone) 29, 44, 51, 79-85
roei (Roe's Abalone) 44, 56
rubra (Blacklip Abalone) 16, 25, 34, 44, 73-77, 131
rubra var. *conicopora* (Brownlip Abalone) 74
scalaris 44
Haliptilon 44
Halisarca laxus 145, **145**
Hapalochlaena
fasciata (Blue-lined Octopus) 92
maculosa (Southern Blue-ringed Octopus) 92
Heliocidaris
erythrogramma (Purple Sea Urchin) 13, 25, 44, 49, 70, 127
tuberculata (Red Sea Urchin) 13
Hemigymnus melapterus (Half and Half Wrasse) 183
Hemiscyllidae 153
Herdmania momus (Sea Squirt) 55
Herklotsichthys castelnaui 212
Heteractis crispa 168
Heteroclinus tristis (Sharp-nose Weedfish) 36
Heterodontidae 151
Heterodontus
galeatus (Crested Horn Shark) 151
portusjacksoni (Port Jackson Shark) 77, 130, 151
Hildenbrandia 66
Hippocampus
abdominalis (Big-belly Seahorse) 46
breviceps (Short-head Seahorse) 46
bargibanti (Pygmy Seahorse) 46
whitei (White's Seahorse) 46
Holopneustes purpurascens 19, 44, **68**, 71
Hormophysa triquetra 48
Hormosira banksii (Neptune's Beads) 37
Hydrurga leptonyx (Leopard Seal) 219
Hyperlophus vittatus 212
Hypoplectrodes 216
mccullochi (Half-banded Seaperch) 171
nigroruber (Banded Sea Perch) 28
Hyporhamphus 212
Idiosepius notoides (Southern Pygmy Squid) 95
Jasus
edwardsii (Southern Rock Lobster) 25, 34, 107–113, 115
lalandii (South African Lobster) 107
novaehollandiae 107
verreauxi (Eastern Rock Lobster) 9, 16, 107, 115–117
Kallymenia 42
Kyphosidae 203
Kyphosus
cornelii (Western Buffalo Bream) 55, 205
sydneyanus (Silver Drummer) 43, 55, 205
vaigiensis (Low-finned Drummer) 205
Labridae 51, 181–87, 189-93, 198
Labroides dimidiatus (Cleaner Wrasse) 185, 215
Laminariales 61
Latrididae 173
Latridopsis forsteri (Bastard Trumpeter) 34
Latris lineata (Striped Trumpeter) 34
Laurencia 42, 44
Lenormandia 45
Leptoichthys fistularius (Brushtail Pipefish) 46
Lessonia corrugata 61
Liagora farinosa 42
Lobophora 56, 66, 69
Lotella rhacina (Large-tooth Beardie) 25, 216
Lotrochota **140**
Lutjanidae 159
Macrocystis 110
angustifolia (String Kelp) 25, 42, 61
pyrifera (Giant Kelp) 31, 36 , 61, 128
Madrella sanguinea 143
Magadena cumingi 45
Manta birostris (Manta Ray) 212
Maoricolpus roseus (New Zealand Screw Shell) 37
Marginaster littoralis 27
Martensia 62
australis 66
Melanthalia 45
Melithaea 43
Meuschenia 196
flavolineata (Yellow-striped Leatherjacket) 198
freycineti (Variable Leatherjacket) 196
hippocrepis (Horseshoe Leatherjacket) 31, 201
trachylepis (Yellow-finned Leatherjacket) 197
venusta (Stars and Stripes Leatherjacket) 197
Microcanthus strigatus (Stripey) 7, 211
Mirounga leonina (Southern Elephant Seal) 219
Mola ramsayi (Short Sunfish) 101
Monacanthidae 27, 51, 19, 195–201, 205, 222
Monacanthus chinensis (Fan-bellied Leatherjacket) 195
Monodactylus argenteus (Silver Batfish) 212
Mopsea 43
Mopsella 43
zimmeri **28, 138**
Mucropetraliella ellerii **142**, 143-44, **144**
Mugil cephalus (Sea Mullet) 161
Mycale **18**
Myriodesma
harveyanum 44
quercifolium 44
Nectocarcinus tuberculosus 43
Nectria 26

Nelusetta 197
ayraudi (Chinaman Leatherjacket) 195
Nemadactylus 174
douglasii (Blue Morwong) 174
macropterus (Jackass Morwong) 174
valenciennesi (Queen Snapper) 174
Neophoca cinerea (Australian Sea Lion) 44, 219
Notolabrus
fucicola (Purple Wrasse) 9, 23, 36, 181
gymnogenis (Crimson-banded Wrasse) 9, 77, 181, 193
inscriptus (Inscribed Wrasse) 181
parilus (Brown-spotted Wrasse) 183
tetricus (Blue-throated Wrasse) 9, 23, 36, 43, 77, 82, 183
Nototodarus gouldii (Gould's Squid) 223
Octopus 88
australis (Hammer Octopus) 92
berrima (Southern Keeled Octopus) 92
bunurong (Southern White-Spot Octopus) 92
kaurna (Southern Sand Octopus) 92
maorum (Maori Octopus) 25, 88
pallidus (Pale Octopus) 92
superciliosus (Frilled Pygmy Octopus) 92
tetricus (Sydney Octopus) 88
warringa (Club Pygmy Octopus) 92
Odacidae 203
Odax
acroptilus (Rainbow Fish) 19, 206
cyanomelas (Herring Cale) 19, 25, 34, 43, 71, 203
pullis (Butterfish) 206
Odontaspididae 154
Odontaspis ferox (Herbst's Nurse Shark) 149
Ophiactis resiliens 25
Ophthalmolepis lineolatus (Maori Wrasse) 181
Orectolobidae 151
Orectolobus
maculatus (Spotted Wobbegong) 153
ornatus (Banded Wobbegong) 154
sp. nov. (Western Wobbegong) 154
Osmundaria 45
Otariidae 219
Pagrus
auratus (Snapper) 159, 201
major (Red Bream) 201
Palmaclathrus 45
stipitatus **48**
Panulirus cygnus (Western Rock Lobster) 5, 51, 107, 119ñ25
Parablennius tasmanianus (Tasmanian Blenny) 25
Paraplesiops meleagris (Western Blue Devil) 25, **55**
Parascyllidae 153
Parascyllium variolatum (Varied Catshark) 25
Parika 197
scaber (Velvet Leatherjacket) 196
Parma 165-71, 203
bicolor (Bicolor Scalyfin) 167, 205
mccullochi (McCulloch's Scalyfin) 167, 205
microlepis (White Ear) 16, 167, 192 , 205, 215
occidentalis (Western Scalyfin) 167, 205
oligolepis (Big-scaled Parma) 165, 205
polylepis (Banded Parma) 9, 165, 205
unifasciata (Girdled Parma) 9, 165, 205
victoriae (Victorian Scalyfin) 25, 31, 167, 205
Patelloida victoriana 36
Patiriella 26, 27
parvivipara 27
regularis (New Zealand Seastar) 27, 37
vivipara 27
Pelsartia humeralis (Sea Trumpeter) 124
Pempheris 212
Pentacerotidae 28
Pentacta ignava 25
Pentagonaster dubeni **viii**
Perithalia caudata 44
Pervagor 197
Petricia vernicina (Velvet Seastar) 26
Peyssonnelia 44
Phacelocarpus 36, 44, 45
Phaeophyta 62
Phocidae 219
Phycodurus eques (Leafy Sea Dragon) 46
Phyllacanthus parvispinus 44
Phyllopteryx taeniolatus (Weedy Sea Dragon) 46
Phyllorhiza punctata (White Spot Jellyfish) 99
Phyllospora comosa (Crayweed) 9, 23, 31, 42, 61, 73, 116
Physalia physalis (Bluebottle) 99
Pictilabrus
brauni 181
laticlavius (Senator Wrasse) 23, 82, 181
viridus **x**
Pinna bicolor (Razor Shell) 142
Plagusia chabrus (Red Rock Crab) 25, 43
Plectaster decanus (Mosaic Seastar) 27
Plectorhinchus flavomaculatus (Gold-spotted Sweetlip) 124
Plocamium 23, 36, 44, 45, 66
Pocillopora damicornis 52
Polysiphonia 171
Pomacentridae 51, 203
Pomacentrus 212
milleri (Miller's Damselfish) 165
Pomatomus saltator (Tailor) 161
Porifera 140
Posidonia 174
Prionace glauca (Blue Shark) 151
Prionurus 203
Psammoperca waigiensis (Sand Bass) 124
Pseudolabrus 181
luculentus (Luculentus Wrasse) 16, 181
parilus (Brown-spotted Wrasse) 124
psittaculus (Rosy Wrasse) 184
Pseudorhiza haekeli (Net-patterned Jellyfish) **104**, **200**
Pterocladia 67
lucida 66
Pteronisis plumacea **32**
Pyura 66
spinifera 19, 143, 144-45, **145**
stolonifera (Cunjevoi) 13, 140, 168
Rhabdosargus sarba (Tarwhine) 55
Rhincodon typus (Whale Shark) 149
Rhodoglossum 206
Rhodophyta 62
Sabella spallanzani (European Fanworm) 49
Sardinops neopilchardus (Pilchard) 6, 212
Sargassum 13, 23, 27, 36, 37, 42, 44, 49, 52, 53, **53**, 55, 56, 57, 61, 66, 69, 70, 71, 74, 195
decurrens 48
fallax 44, 45
heteromorphum 48
lacerifolium 45, 48
sonderi 45
spinuligerum 48
verruculosum 45
Scaberia 49
Scarus ghobban (Blue-barred Orange Parrotfish) 203
Schuettea scalaripinnis (Ladder-finned Pomfret) 212
Scobinichthys 197
granulatus (Rough Leatherjacket) 195
Scorpaena 216
Scorpaenidae 25
Scorpis
aequipinnis (Sea Sweep) 23
lineolata (Silver Sweep) 23, 212
violacea (Blue Sweep) 211
Scyliorhinidae 153
Scytalium 48
Scytothalia 44, 53, 57
dorycarpa 42, 57
Seirococcus 23, 36
axillaris 42
Sepia
apama (Giant Cuttlefish) 88, 95
mestus (Small Reaper Cuttlefish) 95
Sepiadarium austrinum (Southern Bobtail Squid) 95
Sepioloidea lineolata (Striped Pyjama Squid) 95
Sepioteuthis australis (Southern Calamari Squid) 88
Seriola 161
dumerili (Amberjack) 161
hippos (Samson Fish) 161
lalandi (Yellowtail Kingfish) 159, 212
rivoliana (Highfin Amberjack) 161
Seriolella brama (Warehou) 34
Serranidae 181, 189, 198
Smilasterias
multipara (Black and White Seastar) 27
tasmaniae (Tasmanian Seastar) 34
Solieria 42
Sonderopelta 44, 45, 66
coriacea 23
Sparidae 124
Spartina anglica (European Rice Grass) 37
Spirastrella 18
Spratelloides robustus 212
Stegastes
fasciolatus (Pacific Gregory) 165
gascoynei (Gold-belly Gregory) 165
obpreptus (Western Gregory) 205
Sterna anaethetus (Bridled Tern) 55
Stichodactyla gigantea 168
Strongylocentrotus 131
Suezichthys aylingi (Crimson Cleaner Wrasse) 185
Sutorectus tentaculatus (Cobbler Wobbegong) 154
Sycozoa pedunculata 48
Syngnathidae 25
Telesto multiflora 48
Thalassoma
lunare (Moon Wrasse) 181
septemfasciata (Seven-banded Wrasse) 183
Thamnaconus 197
analis (Morse-code Leatherjacket) 201
Thamnoclonium dichotomum 37
Thyrsites atun (Barracouta) 37
Tosia 27
Toxopneustes piledus **126**
Trachinops taeniatus (Eastern Hulafish) 211
Trachurus
declivis (Jack Mackerel) 37, 101, 211, 223
novaezelandiae (Yellowtail) 101
Tripneustes gratilla **128**
Turbo
jourdani 44
torquata (Turban Snail) 19, 44
undulatus (Mottled Green Periwinkle or Warrener) 25, 36, 44
Undaria pinnatifida (Japanese Kelp) **35**, 38, 61
Urolophidae 151
Urolophus circularis **156**
Velella velella (By-The-Wind-Sailor) 105
Vincentia conspersa (Southern Cardinalfish) 25
Zoanthus robustus **58**, **141**
Zonaria 44, 62, 66, 69, 70
diesingiana 19